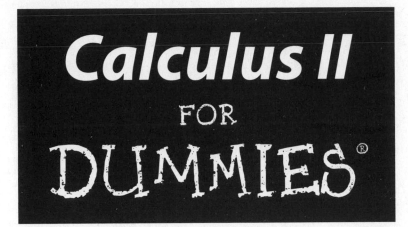

Calculus II
FOR
DUMMIES®

by Mark Zegarelli

WILEY

Wiley Publishing, Inc.

Calculus II For Dummies®

Published by
Wiley Publishing, Inc.
111 River St.
Hoboken, NJ 07030-5774
www.wiley.com

WILEY

About the Author

Mark Zegarelli is the author of *Logic For Dummies* (Wiley), *Basic Math & Pre-Algebra For Dummies* (Wiley), and numerous books of puzzles. He holds degrees in both English and math from Rutgers University, and lives in Long Branch, New Jersey, and San Francisco, California.

Dedication

For my brilliant and beautiful sister, Tami. You are an inspiration.

Author's Acknowledgments

Many thanks for the editorial guidance and wisdom of Lindsay Lefevere, Stephen Clark, and Sarah Faulkner of Wiley Publishing. Thanks also to the Technical Editor, Jeffrey A. Oaks, PhD. Thanks especially to my friend David Nacin, PhD, for his shrewd guidance and technical assistance.

Much love and thanks to my family: Dr. Anthony and Christine Zegarelli, Mary Lou and Alan Cary, Joe and Jasmine Cianflone, and Deseret Moctezuma-Rackham and Janet Rackham. Thanksgiving is at my place this year!

And, as always, thank you to my partner, Mark Dembrowski, for your constant wisdom, support, and love.

Publisher's Acknowledgments

We're proud of this book; please send us your comments through our Dummies online registration form located at www.dummies.com/register/.

Some of the people who helped bring this book to market include the following:

Acquisitions, Editorial, and Media Development

Project Editor: Stephen R. Clark

Acquisitions Editor: Lindsay Sandman Lefevere

Senior Copy Editor: Sarah Faulkner

Editorial Program Coordinator: Erin Calligan Mooney

Technical Editor: Jeffrey A. Oaks, PhD

Editorial Manager: Christine Meloy Beck

Editorial Assistants: Joe Niesen, David Lutton

Cover Photos: Comstock

Cartoons: Rich Tennant (www.the5thwave.com)

Composition Services

Project Coordinator: Katie Key

Layout and Graphics: Carrie A. Cesavice

Proofreaders: Laura Albert, Laura L. Bowman

Indexer: Broccoli Information Management

Special Help

David Nacin, PhD

Publishing and Editorial for Consumer Dummies

Diane Graves Steele, Vice President and Publisher, Consumer Dummies

Joyce Pepple, Acquisitions Director, Consumer Dummies

Kristin A. Cocks, Product Development Director, Consumer Dummies

Michael Spring, Vice President and Publisher, Travel

Kelly Regan, Editorial Director, Travel

Publishing for Technology Dummies

Andy Cummings, Vice President and Publisher, Dummies Technology/General User

Composition Services

Gerry Fahey, Vice President of Production Services

Debbie Stailey, Director of Composition Services

Contents at a Glance

Table of Contents

Introduction

. . . .

Calculus is the great Mount Everest of math. Most of the world is content to just gaze upward at it in awe. But only a few brave souls attempt the ascent.

Or maybe not.

In recent years, calculus has become a required course not only for math, engineering, and physics majors, but also for students of biology, economics, psychology, nursing, and business. Law schools and MBA programs welcome students who've taken calculus because it requires discipline and clarity of mind. Even more and more high schools are encouraging the students to study calculus in preparation for the Advanced Placement (AP) exam.

So, perhaps calculus is more like a well-traveled Vermont mountain, with lots of trails and camping spots, plus a big ski lodge on top. You may need some stamina to conquer it, but with the right guide (this book, for example!), you're not likely to find yourself swallowed up by a snowstorm half a mile from the summit.

About This Book

You, too, can learn calculus. That's what this book is all about. In fact, as you read these words, you may well already be a winner, having passed a course in Calculus I. If so, then congratulations and a nice pat on the back are in order.

Having said that, I want to discuss a few rumors you may have heard about Calculus II:

- ✔ Calculus II is harder than Calculus I.

- ✔ Calculus II is harder, even, than either Calculus III or Differential Equations.

- ✔ Calculus II is more frightening than having your home invaded by zombies in the middle of the night, and will result in emotional trauma requiring years of costly psychotherapy to heal.

Now, I admit that Calculus II is harder than Calculus I. Also, I may as well tell you that many — but not all — math students find it to be harder than the two semesters of math that follow. (Speaking personally, I found Calc II to be easier than Differential Equations.) But I'm holding my ground that the long-term psychological effects of a zombie attack far outweigh those awaiting you in any one-semester math course.

The two main topics of Calculus II are integration and infinite series. *Integration* is the inverse of differentiation, which you study in Calculus I. (For practical purposes, integration is a method for finding the area of unusual geometric shapes.) An *infinite series* is a sum of numbers that goes on forever, like $1 + 2 + 3 + ...$ or $\frac{1}{2} + \frac{1}{4} + \frac{1}{8} + ...$ Roughly speaking, most teachers focus on integration for the first two-thirds of the semester and infinite series for the last third.

This book gives you a solid introduction to what's covered in a college course in Calculus II. You can use it either for self-study or while enrolled in a Calculus II course.

So feel free to jump around. Whenever I cover a topic that requires information from earlier in the book, I refer you to that section in case you want to refresh yourself on the basics.

Here are two pieces of advice for math students — remember them as you read the book:

- **Study a little every day.** I know that students face a great temptation to let a book sit on the shelf until the night before an assignment is due. This is a particularly poor approach for Calc II. Math, like water, tends to seep in slowly and swamp the unwary!

  So, when you receive a homework assignment, read over every problem as soon as you can and try to solve the easy ones. Go back to the harder problems every day, even if it's just to reread and think about them. You'll probably find that over time, even the most opaque problem starts to make sense.

- **Use practice problems for practice.** After you read through an example and think you understand it, copy the problem down on paper, close the book, and try to work it through. If you can get through it from beginning to end, you're ready to move on. If not, go ahead and peek — but then try solving the problem later without peeking. (Remember, on exams, no peeking is allowed!)

Conventions Used in This Book

Throughout the book, I use the following conventions:

- *Italicized* text highlights new words and defined terms.
- **Boldfaced** text indicates keywords in bulleted lists and the action part of numbered steps.
- `Monofont` text highlights Web addresses.
- Angles are measured in radians rather than degrees, unless I specifically state otherwise. See Chapter 2 for a discussion about the advantages of using radians for measuring angles.

What You're Not to Read

All authors believe that each word they write is pure gold, but you don't have to read every word in this book unless you really want to. You can skip over sidebars (those gray shaded boxes) where I go off on a tangent, unless you find that tangent interesting. Also feel free to pass by paragraphs labeled with the Technical Stuff icon.

If you're not taking a class where you'll be tested and graded, you can skip paragraphs labeled with the Tip icon and jump over extended step-by-step examples. However, if you're taking a class, read this material carefully and practice working through examples on your own.

Foolish Assumptions

Not surprisingly, a lot of Calculus II builds on topics introduced Calculus I and Pre-Calculus. So, here are the foolish assumptions I make about you as you begin to read this book:

- If you're a student in a Calculus II course, I assume that you passed Calculus I. (Even if you got a D-minus, your Calc I professor and I agree that you're good to go!)
- If you're studying on your own, I assume that you're at least passably familiar with some of the basics of Calculus I.

I expect that you know some things from Calculus I, but I don't throw you in the deep end of the pool and expect you to swim or drown. Chapter 2 contains a ton of useful math tidbits that you may have missed the first time around. And throughout the book, whenever I introduce a topic that calls for previous knowledge, I point you to an earlier chapter or section so that you can get a refresher.

How This Book Is Organized

This book is organized into six parts, starting you off at the beginning of Calculus II, taking you all the way through the course, and ending with a look at some advanced topics that await you in your further math studies.

Part 1: Introduction to Integration

In Part I, I give you an overview of Calculus II, plus a review of more foundational math concepts.

Chapter 1 introduces the definite integral, a mathematical statement that expresses area. I show you how to formulate and think about an area problem by using the notation of calculus. I also introduce you to the Riemann sum equation for the integral, which provides the definition of the definite integral as a limit. Beyond that, I give you an overview of the entire book

Chapter 2 gives you a need-to-know refresher on Pre-Calculus and Calculus I.

Chapter 3 introduces the indefinite integral as a more general and often more useful way to think about the definite integral.

Part 11: Indefinite Integrals

Part II focuses on a variety of ways to solve indefinite integrals.

Chapter 4 shows you how to solve a limited set of indefinite integrals by using anti-differentiation — that is, by reversing the differentiation process. I show you 17 basic integrals, which mirror the 17 basic derivatives from Calculus I. I also show you a set of important rules for integrating.

Chapter 5 covers variable substitution, which greatly extends the usefulness of anti-differentiation. You discover how to change the variable of a function that you're trying to integrate to make it more manageable by using the integration methods in Chapter 4.

Chapter 6 introduces integration by parts, which allows you to integrate functions by splitting them into two separate factors. I show you how to recognize functions that yield well to this approach. I also show you a handy method — the DI-agonal method — to integrate by parts quickly and easily.

In Chapter 7, I get you up to speed integrating a whole host of trig functions. I show you how to integrate powers of sines and cosines, and then tangents and secants, and finally cotangents and cosecants. Then you put these methods to use in trigonometric substitution.

In Chapter 8, I show you how to use partial fractions as a way to integrate complicated rational functions. As with the other methods in this part of the book, using partial fractions gives you a way to tweak functions that you don't know how to integrate into more manageable ones.

Part III: Intermediate Integration Topics

Part III discusses a variety of intermediate topics, after you have the basics of integration under your belt.

Chapter 9 gives you a variety of fine points to help you solve more complex area problems. You discover how to find unusual areas by piecing together one or more integrals. I show you how to evaluate improper integrals — that is, integrals extending infinitely in one direction. I discuss how the concept of signed area affects the solution to integrals. I show you how to find the average value of a function within an interval. And I give you a formula for finding arc-length, which is the length measured along a curve.

And Chapter 10 adds a dimension, showing you how to use integration to find the surface area and volume of solids. I discuss the meat-slicer method and the shell method for finding solids. I show you how to find both the volume and surface area of revolution. And I show you how to set up more than one integral to calculate more complicated volumes.

Part IV: Infinite Series

In Part IV, I introduce the infinite series — that is, the sum of an infinite number of terms.

Chapter 11 gets you started working with a few basic types of infinite series. I start off by discussing infinite sequences. Then I introduce infinite series, getting you up to speed on expressing a series by using both sigma notation and expanded notation. Then I show you how every series has two associated sequences. To finish up, I introduce you to two common types of series — the geometric series and the p-series — showing you how to recognize and, when possible, evaluate them.

In Chapter 12, I show you a bunch of tests for determining whether a series is convergent or divergent. To begin, I show you the simple but useful nth-term test for divergence. Then I show you two comparison tests — the direct comparison test and the limit comparison test. After that, I introduce you to the more complicated integral, ratio, and root tests. Finally, I discuss alternating series and show you how to test for both absolute and conditional convergence.

And in Chapter 13, the focus is on a particularly useful and expressive type of infinite series called the Taylor series. First, I introduce you to power series. Then I show you how a specific type of power series — the Maclaurin series — can be useful for expressing functions. Finally, I discuss how the Taylor series is a more general version of the Maclaurin series. To finish up, I show you how to calculate the error bounds for Taylor polynomials.

Part V: Advanced Topics

In Part V, I pull out my crystal ball, showing you what lies in the future if you continue your math studies.

In Chapter 14, I give you an overview of Calculus III, also known as multivariable calculus, the study of calculus in three or more dimensions. First, I discuss vectors and show you a few vector calculations. Next, I introduce you to three different three-dimensional (3-D) coordinate systems: 3-D Cartesian coordinates, cylindrical coordinates, and spherical coordinates. Then I discuss functions of several variables, and I show you how to calculate partial derivatives and multiple integrals of these functions.

Chapter 15 focuses on differential equations — that is, equations with derivatives mixed in as variables. I distinguish ordinary differential equations from partial differential equations, and I show you how to recognize the order of a differential equation. I discuss how differential equations arise in science. Finally, I show you how to solve separable differential equations and how to solve linear first-order differential equations.

Part VI: The Part of Tens

Just for fun, Part VI includes a few top-ten lists on a variety of calculus-related topics.

Chapter 16 provides you with ten insights from Calculus II. These insights provide an overview of the book and its most important concepts.

Chapter 17 gives you ten useful test-taking tips. Some of these tips are specific to Calculus II, but many are generally helpful for any test you may face.

Icons Used in This Book

Throughout the book, I use four icons to highlight what's hot and what's not:

This icon points out key ideas that you need to know. Make sure that you understand the ideas before reading on!

Tips are helpful hints that show you the easy way to get things done. Try them out, especially if you're taking a math course.

Warnings flag common errors that you want to avoid. Get clear where these little traps are hiding so that you don't fall in.

This icon points out interesting trivia that you can read or skip over as you like.

Where to Go from Here

You can use this book either for self-study or to help you survive and thrive in a course in Calculus II.

If you're taking a Calculus II course, you may be under pressure to complete a homework assignment or study for an exam. In that case, feel free to skip right to the topic that you need help with. Every section is self-contained, so you can jump right in and use the book as a handy reference. And when I refer to information that I discuss earlier in the book, I give you a brief review and a pointer to the chapter or section where you can get more information if you need it.

If you're studying on your own, I recommend that you begin with Chapter 1, where I give you an overview of the entire book, and read the chapters from beginning to end. Jump over Chapter 2 if you feel confident about your grounding in Calculus I and Pre-Calculus. And, of course, if you're dying to read about a topic that's later in the book, go for it! You can always drop back to an easier chapter if you get lost.

Part I
Introduction to Integration

The 5th Wave

By Rich Tennant

Calculus Overload

©RICH TENNANT

We can't eat the pizza until Lamar determines the relationship of the 3 small wedges to the 2 larger ones.

In this part . . .

1 give you an overview of Calculus II, plus a review of Pre-Calculus and Calculus I. You discover how to measure the areas of weird shapes by using a new tool: the definite integral. I show you the connection between differentiation, which you know from Calculus I, and integration. And you see how this connection provides a useful way to solve area problems.

Chapter 1

An Aerial View of the Area Problem

*H*umans have been measuring the area of shapes for thousands of years. One practical use for this skill is measuring the area of a parcel of land. Measuring the area of a square or a rectangle is simple, so land tends to get divided into these shapes.

Discovering the area of a triangle, circle, or polygon is also easy, but as shapes get more unusual, measuring them gets harder. Although the Greeks were familiar with the conic sections — parabolas, ellipses, and hyperbolas — they couldn't reliably measure shapes with edges based on these figures.

Descartes's invention of analytic geometry — studying lines and curves as equations plotted on a graph — brought great insight into the relationships among the conic sections. But even analytic geometry didn't answer the question of how to measure the area inside a shape that includes a curve.

In this chapter, I show you how *integral calculus* (*integration* for short) developed from attempts to answer this basic question, called the *area problem*. With this introduction to the definite integral, you're ready to look at the practicalities of measuring area. The key to approximating an area that you don't know how to measure is to slice it into shapes that you do know how to measure (for example, rectangles).

Slicing things up is the basis for the *Riemann sum,* which allows you to turn a sequence of closer and closer approximations of a given area into a limit that gives you the exact area that you're seeking. I walk you through a step-by-step process that shows you exactly how the formal definition for the definite integral arises intuitively as you start slicing unruly shapes into nice, crisp rectangles.

Checking out the Area

Finding the area of certain basic shapes — squares, rectangles, triangles, and circles — is easy. But a reliable method for finding the area of shapes containing more esoteric curves eluded mathematicians for centuries. In this section, I give you the basics of how this problem, called the area problem, is formulated in terms of a new concept, the definite integral.

The *definite integral* represents the area on a graph bounded by a function, the *x*-axis, and two vertical lines called the *limits of integration.* Without getting too deep into the computational methods of integration, I give you the basics of how to state the area problem formally in terms of the definite integral.

Comparing classical and analytic geometry

In classical geometry, you discover a variety of simple formulas for finding the area of different shapes. For example, Figure 1-1 shows the formulas for the area of a rectangle, a triangle, and a circle.

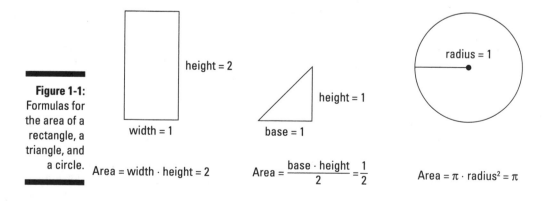

Figure 1-1:
Formulas for the area of a rectangle, a triangle, and a circle.

height = 2

width = 1

Area = width · height = 2

height = 1

base = 1

$$\text{Area} = \frac{\text{base} \cdot \text{height}}{2} = \frac{1}{2}$$

radius = 1

$$\text{Area} = \pi \cdot \text{radius}^2 = \pi$$

Wisdom of the ancients

Long before calculus was invented, the ancient Greek mathematician Archimedes used his *method of exhaustion* to calculate the exact area of a segment of a parabola. Indian mathematicians also developed *quadrature* methods for some difficult shapes before Europeans began their investigations in the 17th century.

These methods anticipated some of the methods of calculus. But before calculus, no single theory could measure the area under arbitrary curves.

When you move on to analytic geometry — geometry on the Cartesian graph — you gain new perspectives on classical geometry. Analytic geometry provides a connection between algebra and classical geometry. You find that circles, squares, and triangles — and many other figures — can be represented by equations or sets of equations, as shown in Figure 1-2.

Figure 1-2:
A rectangle,
a triangle,
and a circle
embedded
on the
graph.

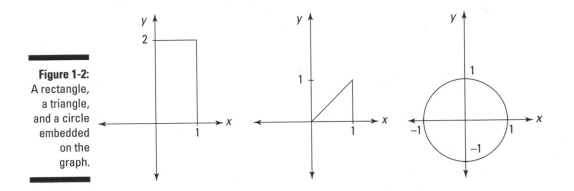

You can still use the trusty old methods of classical geometry to find the areas of these figures. But analytic geometry opens up more possibilities — and more problems.

Discovering a new area of study

Figure 1-3 illustrates three curves that are much easier to study with analytic geometry than with classical geometry: a parabola, an ellipse, and a hyperbola.

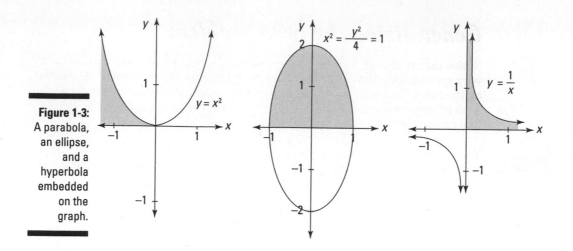

Figure 1-3:
A parabola,
an ellipse,
and a
hyperbola
embedded
on the
graph.

Analytic geometry gives a very detailed account of the connection between algebraic equations and curves on a graph. But analytic geometry doesn't tell you how to find the shaded areas shown in Figure 1-3.

Similarly, Figure 1-4 shows three more equations placed on the graph: a sine curve, an exponential curve, and a logarithmic curve.

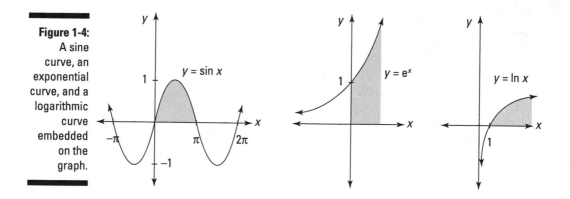

Figure 1-4:
A sine
curve, an
exponential
curve, and a
logarithmic
curve
embedded
on the
graph.

Again, analytic geometry provides a connection between these equations and how they appear as curves on the graph. But it doesn't tell you how to find any of the shaded areas in Figure 1-4.

Generalizing the area problem

Notice that in all the examples in the previous section, I shade each area in a very specific way. Above, the area is bounded by a function. Below, it's bounded by the *x*-axis. And on the left and right sides, the area is bounded by vertical lines (though in some cases, you may not notice these lines because the function crosses the *x*-axis at this point).

You can generalize this problem to study any continuous function. To illustrate this, the shaded region in Figure 1-5 shows the area under the function *f(x)* between the vertical lines *x* = *a* and *x* = *b*.

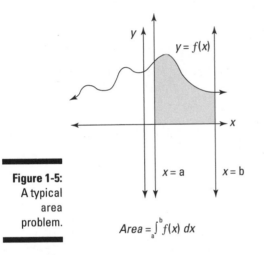

Figure 1-5:
A typical
area
problem.

$$Area = \int_a^b f(x)\ dx$$

The area problem is all about finding the area under a continuous function between two constant values of *x* that are called the *limits of integration,* usually denoted by *a* and *b*.

The limits of integration aren't *limits* in the sense that you learned about in Calculus I. They're simply constants that tell you the width of the area that you're attempting to measure.

In a sense, this formula for the shaded area isn't much different from those that I provide earlier in this chapter. It's just a formula, which means that if you plug in the right numbers and calculate, you get the right answer.

The catch, however, is in the word *calculate*. How exactly do you calculate using this new symbol $\int$? As you may have figured out, the answer is on the cover of this book: *calculus*. To be more specific, *integral calculus* or *integration*.

Most typical Calculus II courses taught at your friendly neighborhood college or university focus on integration — the study of how to solve the area problem. When Calculus II gets confusing (and to be honest, you probably will get confused somewhere along the way), try to relate what you're doing back to this central question: "How does what I'm working on help me find the area under a function?"

Finding definite answers with the definite integral

You may be surprised to find out that you've known how to integrate some functions for years without even knowing it. (Yes, you can know something without knowing that you know it.)

For example, find the rectangular area under the function $y = 2$ between $x = 1$ and $x = 4$, as shown in Figure 1-6.

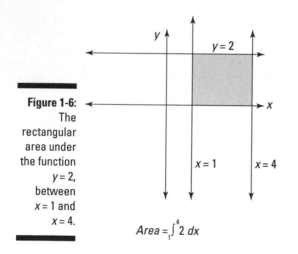

Figure 1-6:
The rectangular area under the function $y = 2$, between $x = 1$ and $x = 4$.

$$Area = \int_1^4 2\,dx$$

This is just a rectangle with a base of 3 and a height of 2, so its area is obviously 6. But this is also an area problem that can be stated in terms of integration as follows:

$$Area = \int_1^4 2\,dx = 6$$

As you can see, the function I'm integrating here is $f(x) = 2$. The limits of integration are 1 and 4 (notice that the greater value goes on top). You already know that the area is 6, so you can solve this calculus problem without resorting to any scary or hairy methods. But, you're still *integrating,* so please pat yourself on the back, because I can't quite reach it from here.

The following expression is called a definite integral:

$$\int_1^4 2\,dx$$

For now, don't spend too much time worrying about the deeper meaning behind the $\int$ symbol or the dx (which you may remember from your fond memories of the differentiating that you did in Calculus I). Just think of $\int$ and dx as notation placed around a function — notation that means *area.*

What's so definite about a definite integral? Two things, really:

- ✓ **You definitely know the limits of integration** (in this case, 1 and 4). Their presence distinguishes a definite integral from an *indefinite integral,* which you find out about in Chapter 3. Definite integrals always include the limits of integration; indefinite integrals never include them.

- ✓ **A definite integral definitely equals a number** (assuming that its limits of integration are also numbers). This number may be simple to find or difficult enough to require a room full of math professors scribbling away with #2 pencils. But, at the end of the day, a number is just a number. And, because a definite integral is a measurement of area, you should expect the answer to be a number.

When the limits of integration *aren't* numbers, a definite integral doesn't necessarily equal a number. For example, a definite integral whose limits of integration are k and $2k$ would most likely equal an algebraic expression that includes k. Similarly, a definite integral whose limits of integration are $\sin\theta$ and $2\sin\theta$ would most likely equal a trig expression that includes θ. To sum up, because a definite integral represents an area, it always equals a number — though you may or may not be able to compute this number.

As another example, find the triangular area under the function $y = x$, between $x = 0$ and $x = 8$, as shown in Figure 1-7.

This time, the shape of the shaded area is a triangle with a base of 8 and a height of 8, so its area is 32 (because the area of a triangle is half the base times the height). But again, this is an area problem that can be stated in terms of integration as follows:

$$Area = \int_0^8 x\,dx = 32$$

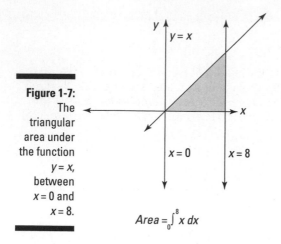

Figure 1-7:
The
triangular
area under
the function
$y = x$,
between
$x = 0$ and
$x = 8$.

$$Area = \int_{0}^{8} x \, dx$$

The function I'm integrating here is $f(x) = x$ and the limits of integration are 0 and 8. Again, you can evaluate this integral with methods from classical and analytic geometry. And again, the definite integral evaluates to a number, which is the area below the function and above the x-axis between $x = 0$ and $x = 8$.

As a final example, find the semicircular area between $x = -4$ and $x = 4$, as shown in Figure 1-8.

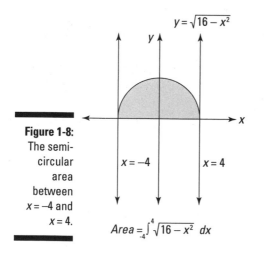

Figure 1-8:
The semi-
circular
area
between
$x = -4$ and
$x = 4$.

$$Area = \int_{-4}^{4} \sqrt{16 - x^2} \, dx$$

First of all, remember from Pre-Calculus how to express the area of a circle with a radius of 4 units:

$$x^2 + y^2 = 16$$

Next, solve this equation for y:

$$y = \pm \sqrt{16 - x^2}$$

A little basic geometry tells you that the area of the whole circle is 16π, so the area of the shaded semicircle is 8π. Even though a circle isn't a function (and remember that integration deals exclusively with *continuous functions!*), the shaded area in this case is beneath the top portion of the circle. The equation for this curve is the following function:

$$y = \sqrt{16 - x^2}$$

So, you can represent this shaded area as a definite integral:

$$Area = \int_{-4}^{4} \sqrt{16 - x^2}\, dx = 8\pi$$

Again, the definite integral evaluates to a number, which is the area under the function between the limits of integration.

Slicing Things Up

One good way of approaching a difficult task — from planning a wedding to climbing Mount Everest — is to break it down into smaller and more manageable pieces.

In this section, I show you the basics of how mathematician Bernhard Riemann used this same type of approach to calculate the definite integral, which I introduce in the previous section "Checking out the Area." Throughout this section I use the example of the area under the function $y = x^2$, between $x = 1$ and $x = 5$. You can find this example in Figure 1-9.

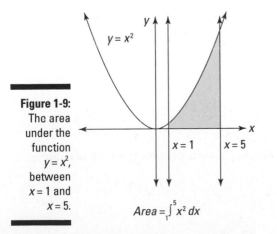

Figure 1-9:
The area under the function $y = x^2$, between $x = 1$ and $x = 5$.

$y = x^2$

$x = 1$ $x = 5$

$Area = \int_{1}^{5} x^2\, dx$

Untangling a hairy problem by using rectangles

The earlier section "Checking out the Area" tells you how to write the definite integral that represents the area of the shaded region in Figure 1-9:

$$\int_1^5 x^2\, dx$$

Unfortunately, this definite integral — unlike those earlier in this chapter — doesn't respond to the methods of classical and analytic geometry that I use to solve the problems earlier in this chapter. (If it did, integrating would be much easier and this book would be a lot thinner!)

Even though you can't solve this definite integral directly (yet!), you can approximate it by slicing the shaded region into two pieces, as shown in Figure 1-10.

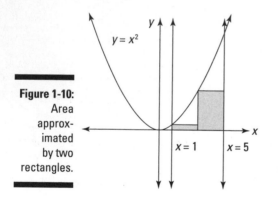

Figure 1-10:
Area approx-
imated
by two
rectangles.

Obviously, the region that's now shaded — it looks roughly like two steps going up but leading nowhere — is less than the area that you're trying to find. Fortunately, these steps do lead someplace, because calculating the area under them is fairly easy.

Each rectangle has a width of 2. The tops of the two rectangles cut across where the function x^2 meets $x = 1$ and $x = 3$, so their heights are 1 and 9, respectively. So, the total area of the two rectangles is 20, because

$$2\,(1) + 2\,(9) = 2\,(1 + 9) = 2\,(10) = 20$$

With this approximation of the area of the original shaded region, here's the conclusion you can draw:

$$\int_1^5 x^2\, dx \approx 20$$

Granted, this is a ballpark approximation with a really big ballpark. But, even a lousy approximation is better than none at all. To get a better approximation, try cutting the figure that you're measuring into a few more slices, as shown in Figure 1-11.

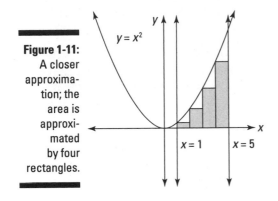

Again, this approximation is going to be less than the actual area that you're seeking. This time, each rectangle has a width of 1. And the tops of the four rectangles cut across where the function x^2 meets $x = 1$, $x = 2$, $x = 3$, and $x = 4$, so their heights are 1, 4, 9, and 16, respectively. So the total area of the four rectangles is 30, because

$$1\,(1) + 1\,(4) + 1\,(9) + 1\,(16) = 1\,(1 + 4 + 9 + 16) = 1\,(30) = 30$$

Therefore, here's a second approximation of the shaded area that you're seeking:

$$\int_1^5 x^2\, dx \approx 30$$

Your intuition probably tells you that your second approximation is better than your first, because slicing the rectangles more thinly allows them to cut in closer to the function. You can verify this intuition by realizing that both 20 and 30 are *less* than the actual area, so whatever this area turns out to be, 30 must be closer to it.

How high is up?

When you're slicing a weird shape into rectangles, finding the width of each rectangle is easy because they're all the same width. You just divide the total width of the area that you're measuring into equal slices.

Finding the height of each individual rectangle, however, requires a bit more work. Start by drawing the horizontal tops of all the rectangles you'll be using. Then, for each rectangle:

1. Locate where the top of the rectangle meets the function.

2. Find the value of x at that point by looking down at the x-axis directly below this point.

3. Get the height of the rectangle by plugging that x-value into the function.

You might imagine that by slicing the area into more rectangles (say 10, or 100, or 1,000,000), you'd get progressively better estimates. And, again, your intuition would be correct: As the number of slices increases, the result approaches 41.3333....

In fact, you may very well decide to write:

$$\int_1^5 x^2\,dx = 41.\overline{33}$$

This, in fact, is the correct answer. But to justify this conclusion, you need a bit more rigor.

Building a formula for finding area

In the previous section, you calculate the areas of two rectangles and four rectangles, respectively, as follows:

$$2\,(1) + 2\,(9) = 2\,(1 + 9) = 20$$
$$1\,(1) + 1\,(4) + 1\,(9) + 1\,(16) = 1\,(1 + 4 + 9 + 16) = 30$$

Each time, you divide the area that you're trying to measure into rectangles that all have the same width. Then, you multiply this width by the sum of the heights of *all* the rectangles. The result is the area of the shaded area.

In general, then, the formula for calculating an area sliced into n rectangles is:

Area of rectangles = $wh_1 + wh_2 + \ldots + wh_n$

In this formula, w is the width of each rectangle and $h_1, h_2, \dots, h_n$, and so forth are the various heights of the rectangles. The width of all the rectangles is the same, so you can simplify this formula as follows:

Area of rectangles = $w\,(h_1 + h_2 + \dots + h_n)$

Remember that as n increases — that is, the more rectangles you draw — the total area of all the rectangles approaches the area of the shape that you're trying to measure.

I hope that you agree that there's nothing terribly tricky about this formula. It's just basic geometry, measuring the area of rectangles by multiplying their width and height. Yet, in the rest of this section, I transform this simple formula into the following formula, called the *Riemann sum formula* for the definite integral:

$$\int_a^b f(x)\,dx = \lim_{n \to \infty} \sum_{i=1}^{n} f\left(x_i^*\right)\left(\frac{b-a}{n}\right)$$

No doubt about it, this formula is eye-glazing. That's why I build it step by step by starting with the simple area formula. This way, you understand completely how all this fancy notation is really just an extension of what you can see for yourself.

If you're sketchy on any of these symbols — such as Σ and the limit — read on, because I explain them as I go along. (For a more thorough review of these symbols, see Chapter 2.)

Approximating the definite integral

Earlier in this chapter I tell you that the definite integral means area. So in transforming the simple formula

Area of rectangles = $w\,(h_1 + h_2 + \dots + h_n)$

the first step is simply to introduce the definite integral:

$$\int_a^b f(x)\,dx \approx w\left(h_1 + h_2 + \dots + h_n\right)$$

As you can see, the = has been changed to $\approx$ — that is, the equation has been demoted to an approximation. This change is appropriate — the definite integral is the *precise* area inside the specified bounds, which the area of the rectangles merely approximates.

Limiting the margin of error

As n increases — that is, the more rectangles you draw — your approximation gets better and better. In other words, as n approaches infinity, the area of the rectangles that you're measuring approaches the area that you're trying to find.

So, you may not be surprised to find that when you express this approximation in terms of a limit, you remove the margin of error and restore the approximation to the status of an equation:

$$\int_a^b f(x)\,dx = \lim_{n \to \infty} w\left(h_1 + h_2 + \ldots + h_n\right)$$

This limit simply states mathematically what I say in the previous section: As *n* approaches infinity, the area of all the rectangles approaches the *exact* area that the definite integral represents.

Widening your understanding of width

The next step is to replace the variable *w*, which stands for the width of each rectangle, with an expression that's more useful.

Remember that the limits of integration tell you the width of the area that you're trying to measure, with *a* as the smaller value and *b* as the greater. So you can write the width of the entire area as *b – a*. And when you divide this area into *n* rectangles, each rectangle has the following width:

$$w = \frac{b - a}{n}$$

Substituting this expression into the approximation results in the following:

$$\int_a^b f(x)\,dx = \lim_{n \to \infty} \frac{b - a}{n}\left(h_1 + h_2 + \ldots + h_n\right)$$

As you can see, all I'm doing here is expressing the variable *w* in terms of *a*, *b*, and *n*.

Summing things up with sigma notation

You may remember that sigma notation — the Greek symbol Σ used in equations — allows you to streamline equations that have long strings of numbers added together. Chapter 2 gives you a review of sigma notation, so check it out if you need a review.

The expression $h_1 + h_2 + \ldots + h_n$ is a great candidate for sigma notation:

$$\sum_{i=1}^{n} h_i = h_1 + h_2 + \ldots + h_n$$

So, in the equation that you're working with, you can make a simple substitution as follows:

$$\int_a^b f(x)\,dx = \lim_{n \to \infty} \frac{b - a}{n} \sum_{i=1}^{n} h_i$$

Now, I tweak this equation by placing $\frac{b-a}{n}$ inside the sigma expression (this is a valid rearrangement, as I explain in Chapter 2):

$$\int_a^b f(x)\,dx = \lim_{n \to \infty} \sum_{i=1}^n h_i \left(\frac{b-a}{n} \right)$$

Heightening the functionality of height

Remember that the variable h_i represents the height of a single rectangle that you're measuring. (The sigma notation takes care of adding up these heights.) The last step is to replace h_i with something more functional. And *functional* is the operative word, because the *function* determines the height of each rectangle.

Here's the short explanation, which I clarify later: The height of each individual rectangle is determined by a value of the function at some value of x lying someplace on that rectangle, so:

$h_i = f(x_i^*)$

The notation x_i^*, which I explain further in "Moving left, right, or center," means something like "an appropriate value of x_i." That is, for each h_i in your sum (h_1, h_2, and so forth) you can replace the variable h_i in the equation for an appropriate value of the function. Here's how this looks:

$$\int_a^b f(x)\,dx = \lim_{n \to \infty} \sum_{i=1}^n f(x_i^*) \left(\frac{b-a}{n} \right)$$

This is the complete Riemann sum formula for the definite integral, so in a sense I'm done. But I still owe you a complete explanation for this last substitution, and here it comes.

Moving left, right, or center

Go back to the example that I start with, and take another look at the way I slice the shaded area into four rectangles in Figure 1-12.

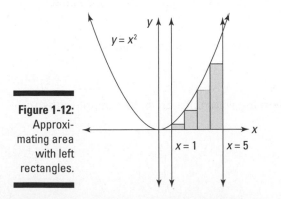

Figure 1-12: Approximating area with left rectangles.

As you can see, the heights of the four rectangles are determined by the value of $f(x)$ when x is equal to 1, 2, 3, and 4, respectively — that is, $f(1)$, $f(2)$, $f(3)$, and $f(4)$. Notice that the upper-left corner of each rectangle touches the function and determines the height of each rectangle.

However, suppose that I draw the rectangles as shown in Figure 1-13.

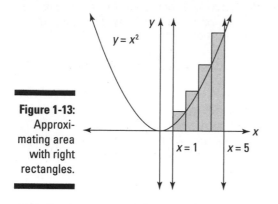

Figure 1-13:
Approximating area with right rectangles.

In this case, the upper-right corner touches the function, so the heights of the four rectangles are $f(2)$, $f(3)$, $f(4)$, and $f(5)$.

Now, suppose that I draw the rectangles as shown in Figure 1-14.

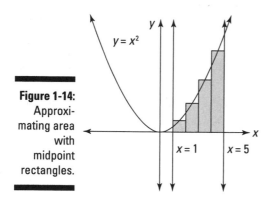

Figure 1-14:
Approximating area with midpoint rectangles.

This time, the midpoint of the top edge of each rectangle touches the function, so the heights of the rectangles are $f(1.5)$, $f(2.5)$, $f(3.5)$, and $f(4.5)$.

It seems that I can draw rectangles at least three different ways to approximate the area that I'm attempting to measure. They all lead to different approximations, so which one leads to the correct answer? The answer is *all of them*.

This surprising answer results from the fact that the equation for the definite integral includes a limit. No matter how you draw the rectangles, as long as the top of each rectangle coincides with the function at one point (at least), the limit smoothes over any discrepancies as n approaches infinity. This slack in the equation shows up as the * in the expression $f(x_i{}^*)$.

For example, in the example that uses four rectangles, the first rectangle is located from $x = 1$ to $x = 2$, so

$$1 \le x_1{}^* \le 2 \quad \text{therefore} \quad 1 \le f(x_1{}^*) \le 4$$

Table 1-2 shows you the range of allowable values for x_i when approximating this area with four rectangles. In each case, you can draw the height of the rectangle on a range of different values of x.

Table 1-2	Allowable Values of $x_i{}^*$ When $n = 4$			
Value of i	*Location of Rectangle*	*Allowable Value of $x_i{}^*$*	*Lowest Value of $f(x_i{}^*)$*	*Highest Value of $f(x_i{}^*)$*
$i = 1$	$x = 1$ to $x = 2$	$1 \le x_1{}^* \le 2$	$f(1) = 1$	$f(2) = 4$
$i = 2$	$x = 2$ to $x = 3$	$2 \le x_2{}^* \le 3$	$f(2) = 4$	$f(3) = 9$
$i = 3$	$x = 3$ to $x = 4$	$3 \le x_3{}^* \le 4$	$f(3) = 9$	$f(4) = 16$
$i = 4$	$x = 4$ to $x = 5$	$4 \le x_1{}^* \le 5$	$f(4) = 16$	$f(5) = 25$

In Chapter 3, I discuss this idea — plus a lot more about the fine points of the formula for the definite integral — in greater detail.

Defining the Indefinite

The Riemann sum formula for the definite integral, which I discuss in the previous section, allows you to calculate areas that you can't calculate by using classical or analytic geometry. The downside of this formula is that it's quite a hairy beast. In Chapter 3, I show you how to use it to calculate area, but most students throw their hands up at this point and say, "There has to be a better way!"

The better way is called the *indefinite integral*. The indefinite integral looks a lot like the definite integral. Compare for yourself:

Definite Integrals   **Indefinite Integrals**

$$\int_{1}^{5} x^2\,dx \qquad\qquad \int x^2\,dx$$

$$\int_{0}^{\pi} \sin x\,dx \qquad\qquad \int \sin x\,dx$$

$$\int_{-1}^{1} e^x\,dx \qquad\qquad \int e^x\,dx$$

Like the definite integral, the indefinite integral is a tool for measuring the area under a function. Unlike it, however, the indefinite integral has no limits of integration, so evaluating it doesn't give you a number. Instead, when you evaluate an indefinite integral, the result is a *function* that you can use to obtain all related definite integrals. Chapter 3 gives you the details of how definite and indefinite integrals are related.

Indefinite integrals provide a convenient way to calculate definite integrals. In fact, the indefinite integral is the *inverse* of the derivative, which you know from Calculus I. (Don't worry if you don't remember all about the derivative — Chapter 2 gives you a thorough review.) By inverse, I mean that the indefinite integral of a function is really the *anti-derivative* of that function. This connection between integration and differentiation is more than just an odd little fact: It's known as the *Fundamental Theorem of Calculus* (FTC).

For example, you know that the derivative of x^2 is $2x$. So, you expect that the anti-derivative — that is, the indefinite integral — of $2x$ is x^2. This is fundamentally correct with one small tweak, as I explain in Chapter 3.

Seeing integration as anti-differentiation allows you to solve tons of integrals without resorting to the Riemann sum formula (I tell you about this in Chapter 4). But integration can still be sticky depending on the function that you're trying to integrate. Mathematicians have developed a wide variety of techniques for evaluating integrals. Some of these methods are variable substitution (see Chapter 5), integration by parts (see Chapter 6), trig substitution (see Chapter 7), and integration by partial fractions (see Chapter 8).

Solving Problems with Integration

After you understand how to describe an area problem by using the definite integral (Part I), and how to calculate integrals (Part II), you're ready to get into action solving a wide range of problems.

Some of these problems know their place and stay in two dimensions. Others rise up and create a revolution in three dimensions. In this section, I give you a taste of these types of problems, with an invitation to check out Part III of this book for a deeper look.

Three types of problems that you're almost sure to find on an exam involve finding the area between curves, the length of a curve, and volume of revolution. I focus on these types of problems and many others in Chapters 9 and 10.

We can work it out: Finding the area between curves

When you know how the definite integral represents the area under a curve, finding the area between curves isn't too difficult. Just figure out how to break the problem into several smaller versions of the basic area problem. For example, suppose that you want to find the area between the function $y = \sin x$ and $y = \cos x$, from $x = 0$ to $x = \frac{\pi}{4}$ — that is, the shaded area A in Figure 1-15.

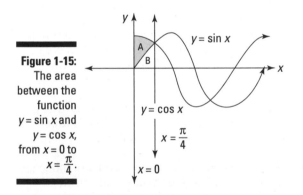

Figure 1-15:
The area between the function $y = \sin x$ and $y = \cos x$, from $x = 0$ to $x = \frac{\pi}{4}$.

In this case, integrating $y = \cos x$ allows you to find the total area A + B. And integrating $y = \sin x$ gives you the area of B. So, you can subtract A + B – B to find the area of A.

For more on how to find an area between curves, flip to Chapter 9.

Walking the long and winding road

Measuring a segment of a straight line or a section of a circle is simple when you're using classical and analytic geometry. But how do you measure a length along an unusual curve produced by a polynomial, exponential, or trig equation?

For example, what's the distance from point A to point B along the curve shown in Figure 1-16?

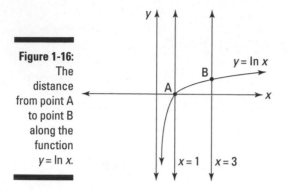

Figure 1-16:
The
distance
from point A
to point B
along the
function
$y = \ln x$.

Once again, integration is your friend. In Chapter 9, I show you how to use integration provides a formula that allows you to measure arc length.

You say you want a revolution

Calculus also allows you to find the volume of unusual shapes. In most cases, calculating volume involves a dimensional leap into *multivariable calculus,* the topic of Calculus III, which I touch upon in Chapter 14. But in a few situations, setting up an integral just right allows you to calculate volume by integrating over a single variable — that is, by using the methods you discover in Calculus II.

Among the trickiest of these problems involves the *solid of revolution* of a curve. In such problems, you're presented with a region under a curve. Then, you imagine the solid that results when you spin this region around the axis, and then you calculate the volume of this solid as seen in Figure 1-17.

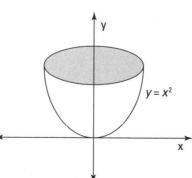

Figure 1-17:
A solid of
revolution
produced by
spinning the
function
$y = x^2$ around
the axis
$x = 0$.

Clearly, you need calculus to find the area of this region. Then you need more calculus and a clear plan of attack to find the volume. I give you all this and more in Chapter 10.

Understanding Infinite Series

The last third of a typical Calculus II course — roughly five weeks — usually focuses on the topic of infinite series. I cover this topic in detail in Part IV. Here's an overview of some of the ideas you find out about there.

Distinguishing sequences and series

A *sequence* is a string of numbers in a determined order. For example:

2, 4, 6, 8, 10, ...

$1, \frac{1}{2}, \frac{1}{4}, \frac{1}{8}, \frac{1}{16}, ...$

$1, \frac{1}{2}, \frac{1}{3}, \frac{1}{4}, \frac{1}{5}, ...$

Sequences can be finite or infinite, but calculus deals well with the infinite, so it should come as no surprise that calculus concerns itself only with *infinite sequences.*

You can turn an infinite sequence into an *infinite series* by changing the commas into plus signs:

2 + 4 + 6 + 8 + 10 + ...

$1 + \frac{1}{2} + \frac{1}{4} + \frac{1}{8} + \frac{1}{16} + ...$

$1 + \frac{1}{2} + \frac{1}{3} + \frac{1}{4} + \frac{1}{5} + ...$

Sigma notation, which I discuss further in Chapter 2, is useful for expressing infinite series more succinctly:

$$\sum_{n=1}^{\infty} 2n$$

$$\sum_{n=1}^{\infty} \left(\frac{1}{2}\right)^n$$

$$\sum_{n=1}^{\infty} \frac{1}{n}$$

Evaluating series

Evaluating an infinite series is often possible. That is, you can find out what all those numbers add up to. For example, here's a solution that should come as no surprise:

$$\sum_{n=1}^{\infty} 2n = 2 + 4 + 6 + 8 + 10 + ... = \infty$$

A helpful way to get a handle on some series is to create a related *sequence of partial sums* — that is, a sequence that includes the first term, the sum of the first two terms, the sum of the first three terms, and so forth. For example, here's a sequence of partial sums for the second series shown earlier:

$$1 = \mathbf{1}$$

$$1 + \frac{1}{2} = \mathbf{1\frac{1}{2}}$$

$$1 + \frac{1}{2} + \frac{1}{4} = \mathbf{1\frac{3}{4}}$$

$$1 + \frac{1}{2} + \frac{1}{4} + \frac{1}{8} = \mathbf{1\frac{7}{8}}$$

$$1 + \frac{1}{2} + \frac{1}{4} + \frac{1}{8} + \frac{1}{16} = \mathbf{1\frac{15}{16}}$$

The resulting sequence of partial sums provides strong evidence of this conclusion:

$$\sum_{n=1}^{\infty} \left(\frac{1}{2}\right)^n = 1 + \frac{1}{2} + \frac{1}{4} + \frac{1}{8} + \frac{1}{16} + ... 1$$

Identifying convergent and divergent series

When a series evaluates to a number — as does $\sum_{n=1}^{\infty} \left(\frac{1}{2}\right)^n$ — it's called a *convergent series*. However, when a series evaluates to infinity — like $\sum_{n=1}^{\infty} 2n$ — it's called a *divergent series.*

Identifying whether a series is convergent or divergent isn't always simple. For example, take another look at the third series I introduce earlier in this section:

$$\sum_{n=1}^{\infty} \frac{1}{n} = 1 + \frac{1}{2} + \frac{1}{3} + \frac{1}{4} + \frac{1}{5} + ... = ?$$

This is called the *harmonic series,* but can you guess by looking at it whether it converges or diverges? (Before you begin adding fractions, let me warn you that the partial sum of the first 10,000 numbers is less than 10.)

An ongoing problem as you study infinite series is deciding whether a given series is convergent or divergent. Chapter 13 gives you a slew of tests to help you find out.

Advancing Forward into Advanced Math

Although it's further along in math than many people dream of going, calculus isn't the end but a beginning. Whether you're enrolled in a Calculus II class or reading on your own, here's a brief overview of some areas of math that lie beyond integration.

Multivariable calculus

Multivariable calculus generalizes differentiation and integration to three dimensions and beyond. Differentiation in more than two dimensions requires *partial derivatives*. Integration in more than two dimensions utilizes *multiple integrals*.

In practice, multivariable calculus as taught in most Calculus III classes is restricted to three dimensions, using three sets of axes and the three variables x, y, and z. I discuss multivariable calculus in more detail in Chapter 14.

Partial derivatives

As you know from Calculus I, a *derivative* is the slope of a curve at a given point on the graph. When you extend the idea of slope to three dimensions, a new set of issues that need to be resolved arises.

For example, suppose that you're standing on the side of a hill that slopes upward. If you draw a line up and down the hill through the point you're standing on, the slope of this line will be steep. But if you draw a line across the hill through the same point, the line will have little or no slope at all. (For this reason, mountain roads tend to cut sideways, winding their way up slowly, rather than going straight up and down.)

So, when you measure slope on a curved surface in three dimensions, you need to take into account not only the *point* where you're measuring the slope but the *direction* in which you're measuring it. Partial derivatives allow you to incorporate this additional information.

Multiple integrals

Earlier in this chapter, you discover that integration allows you to measure the area under a curve. In three dimensions, the analog becomes finding the volume under a curved surface. *Multiple integrals* (integrals nested inside other integrals) allow you to compute such volume.

Differential equations

After multivariable calculus, the next topic most students learn on their precipitous math journey is *differential equations*.

Differential equations arise in many branches of science, including physics, where key concepts such as velocity and acceleration of an object are computed as first and second derivatives. The resulting equations contain hairy combinations of derivatives that are confusing and tricky to solve. For example:

$$F = m \frac{d^2 s}{dt^2}$$

Beyond ordinary differential equations, which include only ordinary derivatives, *partial differential equations* — such as the heat equation or the Laplace equation — include partial derivatives. For example:

$$\nabla^2 V = \frac{\partial^2 V}{\partial x^2} + \frac{\partial^2 V}{\partial y^2} + \frac{\partial^2 V}{\partial z^2} = 0$$

I provide a look at ordinary and partial differential equations in Chapter 15.

Fourier analysis

So much of physics expresses itself in differential equations that finding reliable methods of solving these equations became a pressing need for 19th-century scientists. Mathematician Joseph Fourier met with the greatest success.

Fourier developed a method for expressing every function as the function of an infinite series of sines and cosines. Because trig functions are continuous and infinitely differentiable, Fourier analysis provided a unified approach to solving huge families of differential equations that were previously incalculable.

Numerical analysis

A lot of math is theoretical and ideal: the search for exact answers without regard to practical considerations such as "How long will this problem take to solve?" (If you've ever run out of time on a math exam, you probably know what I'm talking about!)

In contrast, *numerical analysis* is the search for a close-enough answer in a reasonable amount of time.

For example, here's an integral that can't be evaluated:

$$\int e^{x^2} dx$$

But even though you can't solve this integral, you can *approximate* its solution to any degree of accuracy that you desire. And for real-world applications, a good approximation is often acceptable as long as you (or, more likely, a computer) can calculate it in a reasonable amount of time. Such a procedure for approximating the solution to a problem is called an *algorithm*.

Numerical analysis examines algorithms for qualities such as *precision* (the margin of error for an approximation) and *tractability* (how long the calculation takes for a particular level of precision).

Chapter 2

Dispelling Ghosts from the Past: A Review of Pre-Calculus and Calculus I

In This Chapter

▶ Making sense of exponents of 0, negative numbers, and fractions

▶ Graphing common continuous functions and their transformations

▶ Remembering trig identities and sigma notation

▶ Understanding and evaluating limits

▶ Differentiating by using all your favorite rules

▶ Evaluating indeterminate forms of limits with L'Hospital's Rule

Remember Charles Dickens's *A Christmas Carol*? You know, Scrooge and those ghosts from the past. Math can be just like that: All the stuff you thought was dead and buried for years suddenly pays a spooky visit when you least expect it.

This quick review is here to save you from any unnecessary sleepless nights. Before you proceed any further on your calculus quest, make sure that you're on good terms with the information in this chapter.

First I cover all the Pre-Calculus you forgot to remember: polynomials, exponents, graphing functions and their transformations, trig identities, and sigma notation. Then I give you a brief review of Calculus I, focusing on limits and derivatives. I close the chapter with a topic that you may or may not know from Calculus I: L'Hospital's Rule for evaluating indeterminate forms of limits.

If you still feel stumped after you finish this chapter, I recommend that you pick up a copy of *Pre-Calculus For Dummies* by Deborah Rumsey, PhD, or *Calculus For Dummies* by Mark Ryan (both published by Wiley), for a more in-depth review.

Forgotten but Not Gone: A Review of Pre-Calculus

Here's a true story: When I returned to college to study math, my first degree having been in English, it had been a lot of years since I'd taken a math course. I won't mention how many years, but when I confided this number to my first Calculus teacher, she swooned and was revived with smelling salts (okay, I'm exaggerating a little), and then she asked with a concerned look on her face, "Are you sure you're up for this?"

I wasn't sure at all, but I hung in there. Along the way, I kept refining a stack of notes labeled "Brute Memorization" — basically, what you find in this section. Here's what I learned that semester: Whether it's been one year or 20 since you took Pre-Calculus, make sure that you're comfy with this information.

Knowing the facts on factorials

The *factorial* of a positive integer, represented by the symbol !, is that number multiplied by every positive integer less than itself. For example:

$$5! = 5 \cdot 4 \cdot 3 \cdot 2 \cdot 1 = 120$$

Notice that the factorial of every positive number equals that number multiplied by the next-lowest factorial. For example:

$$6! = 6 \, (5!)$$

Generally speaking, then, the following equality is true:

$$(x + 1)! = (x + 1) \, x!$$

This equality provides the rationale for the odd-looking convention that $0! = 1$:

$$(0 + 1)! = (0 + 1) \, 0!$$
$$1! = (1) \, 0!$$
$$1 = 0!$$

When factorials show up in fractions (as they do in Chapters 12 and 13), you can usually do a lot of cancellation that makes them simpler to work with. For example:

$$\frac{3!}{5!} = \frac{(3 \cdot 2 \cdot 1)}{(5 \cdot 4 \cdot 3 \cdot 2 \cdot 1)} = \frac{1}{(5 \cdot 4)} = \frac{1}{20}$$

Even when a fraction includes factorials with variables, you can usually simplify it. For example:

$$\frac{(x+1)!}{x!} = \frac{(x+1)\,x!}{x!} = x+1$$

Polishing off polynomials

A polynomial is any function of the following form:

$$f(x) = a_n x^n + a_{n-1} x^{n-1} + a_{n-2} x^{n-2} + \ldots + a_1 x + a_0$$

Note that every term in a polynomial is x raised to the power of a nonnegative integer, multiplied by a real-number coefficient. Here are a few examples of polynomials:

$$f(x) = x^3 - 4x^2 + 2x - 5$$
$$f(x) = x^{12} - \frac{3}{4}x^7 + 100x - \pi$$
$$f(x) = (x^2 + 8)(x - 6)^3$$

Note that in the last example, multiplying the right side of the equation will change the polynomial to a more recognizable form.

Polynomials enjoy a special status in math because they're particularly easy to work with. For example, you can find the value of $f(x)$ for any x value by plugging this value into the polynomial. Furthermore, polynomials are also easy to differentiate and integrate. Knowing how to recognize polynomials when you see them will make your life in any math course a whole lot easier.

Powering through powers (exponents)

Remember when you found out that any number (except 0) raised to the power of 0 equals 1? That is:

$$n^0 = 1 \ (\text{for all } n \neq 0)$$

It just seemed weird, didn't it? But when you asked your teacher why, I suspect you got an answer that sounded something like "That's just how mathematicians define it." Not a very satisfying answer, is it?

However if you're absolutely dying to know why (or if you're even mildly curious about it), the answer lies in number patterns.

For starters, suppose that $n = 2$. Table 2-1 is a simple chart that encapsulates information you already know.

Table 2-1			Positive Integer Exponents of 2					
x	1	2	3	4	5	6	7	8
2^x	2	4	8	16	32	64	128	256

As you can see, as x increases by 1, 2^x doubles. So, as x decreases by 1, 2^x is halved. You don't need rocket science to figure out what happens when $x = 0$. Table 2-2 shows you what happens.

Table 2-2			Nonnegative Integer Exponents of 2					
x	**0**	1	2	3	4	5	6	7	8
2^x	**1**	2	4	8	16	32	64	128	256

This chart provides a simple rationale of why $2^0 = 1$. The same reasoning works for all other real values of n (except 0). Furthermore, Table 2-3 shows you what happens when you continue the pattern into negative values of x.

Table 2-3			Positive and Negative Integer Exponents of 2					
x	−4	−3	−2	−1	0	1	2	3	4
n^x	$\frac{1}{16}$	$\frac{1}{8}$	$\frac{1}{4}$	$\frac{1}{2}$	1	2	4	8	16

As the table shows, $2^{-x} = \frac{1}{2^x}$. This pattern also holds for all real, nonzero values of n, so

$$n^{-x} = \frac{1}{n^x}$$

Notice from this table that the following rule holds:

$$n^a n^b = n^{a+b}$$

For example:

$$2^3 \cdot 2^4 = 2^{3+4} = 2^7 = 128$$

This rule allows you to evaluate fractional exponents as roots. For example:

$$2^{\frac{1}{2}} \cdot 2^{\frac{1}{2}} = 2^{\left(\frac{1}{2} + \frac{1}{2}\right)} = 2^{1} = 2 \qquad \text{so } 2^{\frac{1}{2}} = \sqrt{2}$$

You can generalize this rule for all bases and fractional exponents as follows:

$$n^{\frac{a}{b}} = \sqrt[b]{n^{a}}$$

Plotting these values for x and $f(x) = 2^x$ onto a graph provides an even deeper understanding (check out Figure 2-1):

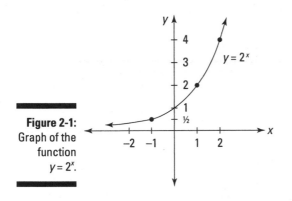

Figure 2-1:
Graph of the function
$y = 2^x$.

In fact, assuming the continuity of the exponential curve even provides a rationale (or, I suppose, an *irrationale*) for calculating a number raised to an irrational exponent. This calculation is beyond the scope of this book, but it's a problem in numerical analysis, a topic that I discuss briefly in Chapter 1.

Noting trig notation

Trigonometry is a big and important subject in Calculus II. I can't cover everything you need to know about trig here. For more detailed information on trig, see *Trigonometry For Dummies* by Mary Jane Sterling (Wiley). But I do want to spend a moment on one aspect of trig notation to clear up any confusion you may have.

When you see the notation

$2 \cos x$

remember that this means 2 (cos x). So, to evaluate this function for $x = \pi$, evaluate the inner function cos x first, and then multiply the result by 2:

$2 \cos \pi = 2 \cdot -1 = -2$

On the other hand, the notation

cos 2*x*

means cos (2*x*). For example, to evaluate this function for *x* = 0, evaluate the inner function 2*x* first, and then take the cosine of the result:

cos (2 · 0) = cos 0 = 1

Finally (and make sure that you understand this one!), the notation

cos^2 *x*

means (cos *x*)^2. In other words, to evaluate this function for *x* = π, evaluate the inner function cos *x* first, and then take the square of the result:

cos^2 π = (cos π)^2 = (–1)^2 = 1

Getting clear on how to evaluate trig functions really pays off when you're applying the Chain Rule (which I discuss later in this chapter) and when integrating trig functions (which I focus on in Chapter 7).

Figuring the angles with radians

When you first discovered trigonometry, you probably used degrees because they were familiar from geometry. Along the way, you were introduced to radians, and forced to do a bunch of conversions between degrees and radians, and then in the next chapter you went back to using degrees.

Degrees are great for certain trig applications, such as land surveying. But for math, radians are the right tool for the job. In contrast, degrees are awkward to work with.

For example, consider the expression sin 1,260°. You probably can't tell just from looking at this expression that it evaluates to 0, because 1,260° is a multiple of 180°.

In contrast, you can tell immediately that the equivalent expression sin7π is a multiple of π. And as an added bonus, when you work with radians, the numbers tend to be smaller and you don't have to add the degree symbol (°).

You don't need to worry about calculating conversions between degrees and radians. Just make sure that you know the most common angles in both degrees and radians. Figure 2-2 shows you some common angles.

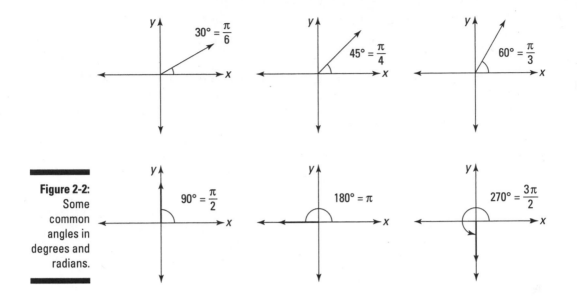

Figure 2-2:
Some common angles in degrees and radians.

Radians are the basis of polar coordinates, which I discuss later in this section.

Graphing common functions

You should be familiar with how certain common functions look when drawn on a graph. In this section, I show you the most common graphs of functions. These functions are all continuous, so they're integrative at all real values of x.

Linear and polynomial functions

Figure 2-3 shows three simple functions.

Figure 2-3:
Graphs of two linear functions $y = n$ and $y = x$ and the absolute value function $y = |x|$.

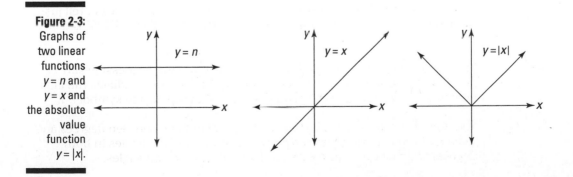

Figure 2-4 includes a few basic polynomial functions.

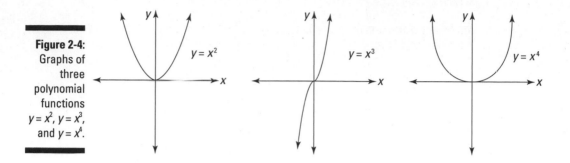

Figure 2-4:
Graphs of three polynomial functions $y = x^2$, $y = x^3$, and $y = x^4$.

Exponential and logarithmic functions

Here are some *exponential functions* with whole number bases:

$$y = 2^x$$

$$y = 3^x$$

$$y = 10^x$$

Notice that for every positive base, the exponential function

- Crosses the y-axis at $x = 1$
- Explodes to infinity as x increases (that is, it has an unbounded y value)
- Approaches $y = 0$ as x decreases (that is, in the negative direction the x-axis is an asymptote)

The most important exponential function is e^x. See Figure 2-5 for a graph of this function.

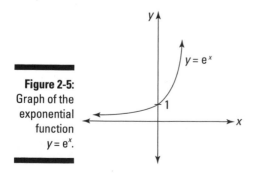

Figure 2-5:
Graph of the exponential function $y = e^x$.

The unique feature of this exponential function is that at every value of x, its slope is e^x. That is, this function is its own derivative (see "Recent Memories: A Review of Calculus I" later in this chapter for more on derivatives).

Another important function is the *logarithmic function* (also called the *natural log function*). Figure 2-6 is a graph of the logarithmic function $y = \ln x$.

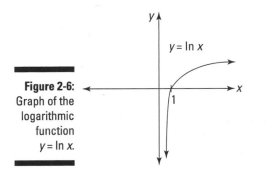

Figure 2-6:
Graph of the logarithmic function
$y = \ln x$.

Notice that this function is the reflection of e^x along the diagonal line $y = x$. So the log function does the following:

✔ Crosses the x-axis at $x = 1$

✔ Explodes to infinity as x increases (that is, it has an unbounded y value), though more slowly than any exponential function

✔ Produces a y value that approaches $-\infty$ as x approaches 0 from the right

Furthermore, the domain of the log functions includes only positive values. That is, inputting a nonpositive value to the log function is a big no-no, on par with placing 0 in the denominator of a fraction or a negative value inside a square root.

For this reason, functions placed inside the log function often get "pretreated" with the absolute value operator. For example:

$$y = \ln |x^3|$$

You can bring an exponent outside of a natural log and make it a coefficient, as follows:

$$\ln (a^b) = b \ln a$$

Trigonometric functions

The two most important graphs of trig functions are the sine and cosine. See Figure 2-7 for graphs of these functions.

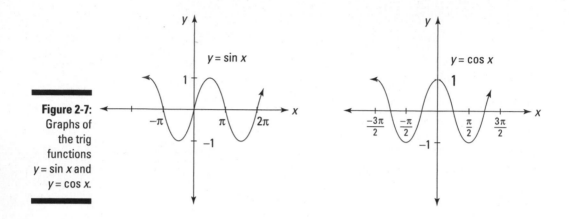

Figure 2-7: Graphs of the trig functions $y = \sin x$ and $y = \cos x$.

Note that the x values of these two graphs are typically marked off in multiples of π. Each of these functions has a period of 2π. In other words, it repeats its values after 2π units. And each has a maximum value of 1 and a minimum value of -1.

Remember that the sine function

- Crosses the origin
- Rises to a value of 1 at $x = \dfrac{\pi}{2}$
- Crosses the x-axis at all multiples of π

Remember that the cosine function

- Has a value of 1 at $x = 0$
- Drops to a value of 0 at $x = \dfrac{\pi}{2}$
- Crosses the x-axis at $\dfrac{3\pi}{2}$, $\dfrac{5\pi}{2}$, $\dfrac{7\pi}{2}$, and so on

The graphs of other trig functions are also worth knowing. Figure 2-8 shows graphs of the trig functions $y = \tan x$, $y = \cot x$, $y = \sec x$, and $y = \csc x$.

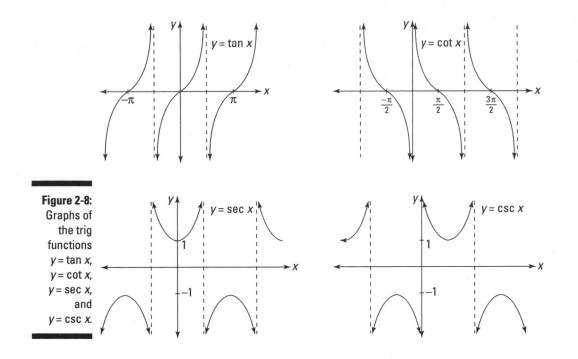

Figure 2-8:
Graphs of
the trig
functions
$y = \tan x$,
$y = \cot x$,
$y = \sec x$,
and
$y = \csc x$.

Asymptotes

An asymptote is any straight line on a graph that a curve approaches but doesn't touch. It's usually represented on a graph as a dashed line. For example, all four graphs in Figure 2-8 have vertical asymptotes.

Depending on the curve, an asymptote can run in any direction, including diagonally. When you're working with functions, however, horizontal and vertical asymptotes are more common.

Transforming continuous functions

When you know how to graph the most common functions, you can transform them by using a few simple tricks, as I show you in Table 2-4.

Table 2-4	Five Vertical and Five Horizontal Transformations of Functions		
Axis	**Direction**	**Transformation**	**Example**
y-axis (vertical)	Shift Up	$y = f(x) + n$	$y = e^x + 1$
	Shift Down	$y = f(x) - n$	$y = x^3 - 2$
	Expand	$y = nf(x)$	$y = 5 \sec x$
	Contract	$y = \dfrac{f(x)}{n}$	$y = \dfrac{\sin x}{10}$
	Reflect	$y = -f(x)$	$y = -(\ln x)$
x-axis (horizontal)	Shift Right	$y = f(x - n)$	$y = e^{x-2}$
	Shift Left	$y = f(x + n)$	$y = (x + 4)^3$
	Expand	$y = f\left(\dfrac{x}{n}\right)$	$y = \sec \dfrac{x}{3}$
	Contract	$y = f(nx)$	$y = \sin(\pi x)$
	Reflect	$y = f(-x)$	$y = e^{-x}$

The vertical transformations are intuitive — that is, they take the function in the direction that you'd probably expect. For example, adding a constant shifts the function up and subtracting a constant shifts it down.

In contrast, the horizontal transformations are counterintuitive — that is, they take the function in the direction that you probably wouldn't expect. For example, adding a constant shifts the function left and subtracting a constant shifts it right.

Identifying some important trig identities

Memorizing trig identities is like packing for a camping trip.

When you're backpacking into the wilderness, there's a limit to what you can comfortably carry, so you should probably leave your pogo stick and your 30-pound dumbbells at home. At the same time, you don't want to find yourself miles from civilization without food, a tent, and a first-aid kit.

I know that committing trig identities to memory registers on the Fun Meter someplace between alphabetizing your spice rack and vacuuming the lint

How to avoid an identity crisis

Most students remember the first square identity without trouble:

$$\sin^2 x + \cos^2 x = 1$$

If you're worried that you might forget the other two square identities just when you need them most, don't despair. An easy way to remember them is to divide every term in the first square identity by $\sin^2 x$ and $\cos^2 x$:

$$\frac{\sin^2 x}{\cos^2 x} + \frac{\cos^2 x}{\cos^2 x} = \frac{1}{\cos^2 x}$$

$$\frac{\sin^2 x}{\sin^2 x} + \frac{\cos^2 x}{\sin^2 x} = \frac{1}{\sin^2 x}$$

Now, simplify these equations using the Basic Five trig identities:

$$1 + \tan^2 x = \sec^2 x$$

$$1 + \cot^2 x = \csc^2 x$$

filter on your dryer. But knowing a few important trig identities can be a life-saver when you're lost out on the misty calculus trails, so I recommend that you take a few along with you. (It's nice when the metaphor really holds up, isn't it?)

For starters, here are the three *inverse identities,* which you probably know already:

$$\sin x = \frac{1}{\csc x}$$

$$\cos x = \frac{1}{\sec x}$$

$$\tan x = \frac{1}{\cot x}$$

You also need these two important identities:

$$\tan x = \frac{\sin x}{\cos x}$$

$$\cot x = \frac{\cos x}{\sin x}$$

I call these the Basic Five trig identities. By using them, you can express *any* trig expression in terms of sines and cosines. Less obviously, you can also express any trig expression in terms of tangents and secants (try it!). Both of these facts are useful in Chapter 7, when I discuss trig integration.

Equally indispensable are the three *square identities.* Most students remember the first and forget about the other two, but you need to know them all:

$$\sin^2 x + \cos^2 x = 1$$

$$1 + \tan^2 x = \sec^2 x$$

$$1 + \cot^2 x = \csc^2 x$$

You also don't want to be seen in public without the two *half-angle identities:*

$$\sin^2 x = \frac{1 - \cos 2x}{2}$$

$$\cos^2 x = \frac{1 + \cos 2x}{2}$$

Finally, you can't live without the *double-angle identities for sines:*

$$\sin 2x = 2 \sin x \cos x$$

Beyond these, if you have a little spare time, you can include these *double-angle identities for cosines and tangents:*

$$\cos 2x = \cos^2 x - \sin^2 x = 2 \cos^2 x - 1 = 1 - 2 \sin^2 x$$

$$\tan 2x = \frac{2 \tan x}{1 - \tan^2 x}$$

Polar coordinates

Polar coordinates are an alternative to the Cartesian coordinate system. As with Cartesian coordinates, polar coordinates assign an ordered pair of values to every point on the plane. Unlike Cartesian coordinates, however, these values aren't (x, y), but rather (r, θ).

✔ The value r is the distance to the origin.

✔ The value θ is the angular distance from the polar axis, which corresponds to the positive x-axis in Cartesian coordinates. (Angular distance is always measured counterclockwise.)

Figure 2-9 shows how to plot points in polar coordinates. For example:

✔ To plot the point $(3, \frac{\pi}{4})$, travel 3 units from the origin on the polar axis, and then arc $\frac{\pi}{4}$ (equivalent to 45°) counterclockwise.

✔ To plot $(4, \frac{5\pi}{6})$, travel 4 units from the origin on the polar axis, and then arc $\frac{5\pi}{6}$ units (equivalent to 150°) counterclockwise.

✔ To plot the point $(2, \frac{3\pi}{2})$, travel 2 units from the origin on the polar axis, and then arc $\frac{3\pi}{2}$ units (equivalent to 270°) counterclockwise.

Polar coordinates allow you to plot certain shapes on the graph more simply than Cartesian coordinates. For example, here's the equation for a 3-unit circle centered at the origin in both Cartesian and polar coordinates:

$$y = \pm \sqrt{x^2 - 9} \qquad r = 3$$

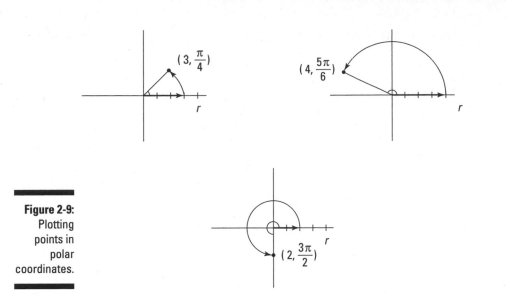

Figure 2-9:
Plotting
points in
polar
coordinates.

Some problems that would be difficult to solve expressed in terms of Cartesian variables (x and y) become much simpler when expressed in terms of polar variables (r and θ). To convert Cartesian variables to polar, use the following formulas:

$$x = r \cos \theta \qquad y = r \sin \theta$$

To convert polar variables to Cartesian, use this formula:

$$r = \pm \sqrt{x^2 + y^2} \qquad \theta = \arctan\left(\frac{y}{x}\right)$$

Polar coordinates are the basis of two alternative 3-D coordinate systems: cylindrical coordinates and spherical coordinates. See Chapter 14 for a look at these two systems.

Summing up sigma notation

Mathematicians just love sigma notation (Σ) for two reasons. First, it provides a convenient way to express a long or even infinite series. But even more important, it looks really cool and scary, which frightens nonmathematicians into revering mathematicians and paying them more money.

However, when you get right down to it, Σ is just fancy notation for adding, and even your little brother isn't afraid of adding, so why should you be?

For example, suppose that you want to add the even numbers from 2 to 10. Of course, you can write this expression and its solution this way:

$$2 + 4 + 6 + 8 + 10 = 30$$

Or you can write the same expression by using sigma notation:

$$\sum_{n=1}^{5} 2n$$

Here, n is the *variable of summation* — that is, the variable that you plug values into and then add them up. Below the Σ, you're given the starting value of n (1) and above it the ending value (5). So here's how to expand the notation:

$$\sum_{n=1}^{5} 2n = 2(1) + 2(2) + 2(3) + 2(4) + 2(5) = 30$$

You can also use sigma notation to stand for the sum of an infinite number of values — that is, an *infinite series.* For example, here's how to add up all the positive square numbers:

$$\sum_{n=1}^{\infty} n^2$$

This compact expression can be expanded as follows:

$$= 1^2 + 2^2 + 3^2 + 4^2 + \ldots = 1 + 4 + 9 + 16 + \ldots$$

This sum is, of course, infinite. But not all infinite series behave in this way. In some cases, an infinite series equals a number. For example:

$$\sum_{n=0}^{\infty} \left(\frac{1}{2}\right)^n$$

This series expands and evaluates as follows:

$$1 + \frac{1}{2} + \frac{1}{4} + \frac{1}{8} + \ldots = 2$$

When a series evaluates to a number, the series is *convergent.* When a series isn't convergent, it's *divergent.* You find out all about divergent and convergent series in Chapter 12.

Recent Memories: A Review of Calculus 1

Integration is the study of how to solve a single problem — the area problem. Similarly, differentiation, which is the focus of Calculus I, is the study of how to solve the *tangent problem:* how to find the slope of the tangent line at any point on a curve. In this section, I review the highlights of Calculus I. For a more thorough review, please see *Calculus For Dummies* by Mark Ryan (Wiley).

Knowing your limits

An important thread that runs through Calculus I is the concept of a *limit*. Limits are also important in Calculus II. In this section, I give you a review of everything you need to remember but may have forgotten about limits.

Telling functions and limits apart

A function provides a link between two variables: the independent variable (usually x) and the dependent variable (usually y). A function tells you the value y when x takes on a specific value. For example, here's a function:

$$y = x^2$$

In this case, when x takes a value of 2, the value of y is 4.

In contrast, a limit tells you what happens to y as x approaches a certain number without actually reaching it. For example, suppose that you're working with the function $y = x^2$ and want to know the limit of this function as x approaches 2. The notation to express this idea is as follows:

$$\lim_{x \to 2} x^2$$

You can get a sense of what this limit equals by plugging successively closer approximations of 2 into the function (see Table 2-5).

Table 2-5		Approximating $\lim_{x \to 2} x^2$			
x	1.7	1.8	1.9	1.99	1.999	1.9999
y	2.69	3.24	3.61	3.9601	3.996001	3.99960001

This table provides strong evidence that the limit evaluates to 4. That is:

$$\lim_{x \to 2} x^2 = 4$$

Remember that this limit tells you nothing about what the function actually equals when $x = 2$. It tells you only that as x approaches 2, the value of the function gets closer and closer to 2. In this case, because the function and the limit are equal, the function is *continuous* at this point.

Evaluating limits

Evaluating a limit means either finding the value of the limit or showing that the limit doesn't exist.

You can evaluate many limits by replacing the limit variable with the number that it approaches. For example:

$$\lim_{x \to 4} \frac{x^2}{2x} = \frac{4^2}{2 \cdot 4} = \frac{16}{8} = 1$$

Sometimes this replacement shows you that a limit doesn't exist. For example:

$$\lim_{x \to \infty} x = \infty$$

When you find that a limit appears to equal either ∞ or $-\infty$, the limit *does not exist (DNE)*. DNE is a perfectly good way to complete the evaluation of a limit.

Some replacements lead to apparently untenable situations, such as division by zero. For example:

$$\lim_{x \to 0} \frac{e^x}{x} = \frac{e^0}{0} = \frac{1}{0}$$

This looks like a dead end, because division by zero is undefined. But, in fact, you can actually get an answer to this problem. Remember that this limit tells you nothing about what happens when x actually equals 0, only what happens as x *approaches* 0: The denominator shrinks toward 0, while the numerator never falls below 1, so the value fraction becomes indefinitely large. Therefore:

$$\lim_{x \to 0} \frac{e^x}{x} \text{ Does Not Exist (DNE)}$$

Here's another example:

$$\lim_{x \to \infty} \frac{1,000,000}{x} = \frac{1,000,000}{\infty}$$

This is another apparent dead end, because ∞ isn't really a number, so how can it be the denominator of a fraction? Again, the limit saves the day. It doesn't tell you what happens when x actually equals ∞ (if such a thing were possible), only what happens as x *approaches* ∞. In this case, the denominator becomes indefinitely large while the numerator remains constant, so:

$$\lim_{x \to \infty} \frac{1,000,000}{x} = 0$$

Some limits are more difficult to evaluate because they're one of several *indeterminate forms.* The best way to solve them is to use L'Hospital's Rule, which I discuss in detail at the end of this chapter.

Hitting the slopes with derivatives

The *derivative* at a given point on a function is the slope of the tangent line to that function at that point. The derivative of a function provides a "slope map" of that function.

The best way to compare a function with its derivative is by lining them up vertically (see Figure 2-10 for an example).

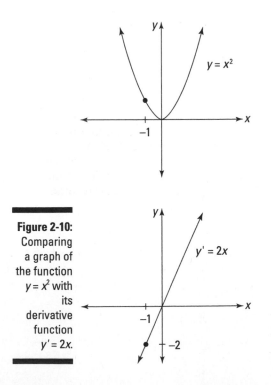

Figure 2-10:
Comparing a graph of the function $y = x^2$ with its derivative function $y' = 2x$.

Looking at the top graph, you can see that when $x = 0$, the slope of the function $y = x^2$ is 0 — that is, no slope. The bottom graph verifies this because at $x = 0$, the derivative function $y = 2x$ is also 0.

You probably can't tell, however, what the slope of the top graph is at $x = -1$. To find out, look at the bottom graph and notice that at $x = -1$, the derivative function equals -2, so -2 is also the slope of the top graph at this point. Similarly, the derivative function tells you the slope at every point on the original function.

Referring to the limit formula for derivatives

In Calculus I, you develop two formulas for the derivative of a function. These formulas are both based on limits, and they're both equally valid:

$$f'(x) = \lim_{h \to 0} \frac{f(x+h) - f(x)}{h}$$

$$f'(x) = \lim_{x \to a} \frac{f(x) - f(a)}{x - a}$$

You probably won't need to refer to these formulas much as you study Calculus II. Still, please keep in mind that the official definition of a function's derivative is always cast in terms of a limit.

For a more detailed look at how these formulas are developed, see *Calculus For Dummies* by Mark Ryan (Wiley).

Knowing two notations for derivatives

Students often find the notation for derivatives — especially Leibniz notation $\frac{d}{dx}$ — confusing. To make things simple, think of this notation as a *unary operator* that works in a similar way to a minus sign.

A minus sign attaches to the front of an expression, changing the value of that expression to its negative. Evaluating the effect of this sign on the expression is called distribution, which produces a new but equivalent expression. For example:

$$-(x^2 + 4x - 5) = -x^2 - 4x + 5$$

Similarly, the notation $\frac{d}{dx}$ attaches to the front of an expression, changing the value of that expression to its *derivative*. Evaluating the effect of this notation on the expression is called *differentiation,* which also produces a new but equivalent expression. For example:

$$\frac{d}{dx}(x^2 + 4x - 5) = 2x + 4$$

The basic notation remains the same even when an expression is recast as a function. For example, given the function $y = f(x) = x^2 + 4x - 5$, here's how you differentiate:

$$\frac{dy}{dx} = \frac{d}{dx}f(x) = 2x + 4$$

The notation $\frac{dy}{dx}$, which means "the change in y as x changes," was first used by Gottfried Leibniz, one of the two inventors of calculus (the other inventor was Isaac Newton). An advantage of Leibniz notation is that it explicitly tells you the variable over which you're differentiating — in this case, x. When this information is easily understood in context, a shorter notation is also available:

$$y' = f'(x) = 2x + 4$$

You should be comfortable with both of these forms of notation. I use them interchangeably throughout this book.

Understanding differentiation

Differentiation — the calculation of derivatives — is the central topic of Calculus I and makes an encore appearance in Calculus II.

In this section, I give you a refresher on some of the key topics of differentiation. In particular, the 17 need-to-know derivatives are here and, for your convenience, in the Cheat Sheet just inside the front cover of this book. And if you're shaky on the Chain Rule, I offer you a clear explanation that gets you up to speed.

Memorizing key derivatives

The derivative of any constant is always 0:

$$\frac{d}{dx}n = 0$$

The derivative of the variable by which you're differentiating (in most cases, x) is 1:

$$\frac{d}{dx}x = 1$$

Here are three more derivatives that are important to remember:

$$\frac{d}{dx} e^x = e^x$$

$$\frac{d}{dx} n^x = n^x \ln n$$

$$\frac{d}{dx} \ln x = \frac{1}{x}$$

You need to know each of these derivatives as you move on in your study of calculus.

Derivatives of the trig functions

The derivatives of the six trig functions are as follows:

$$\frac{d}{dx} \sin x = \cos x$$

$$\frac{d}{dx} \cos x = -\sin x$$

$$\frac{d}{dx} \tan x = \sec^2 x$$

$$\frac{d}{dx} \cot x = -\csc^2 x$$

$$\frac{d}{dx} \sec x = \sec x \tan x$$

$$\frac{d}{dx} \csc x = -\csc x \cot x$$

You need to know them all by heart.

Derivatives of the inverse trig functions

Two notations are commonly used for inverse trig functions. One is the addition of $^{-1}$ to the function: $\sin^{-1}$, $\cos^{-1}$, and so forth. The second is the addition of *arc* to the function: arcsin, arccos, and so forth. They both mean the same thing, but I prefer the *arc* notation, because it's less likely to be mistaken for an exponent.

I know that asking you to memorize these functions seems like a cruel joke. But you really need them when you get to trig substitution in Chapter 7, so at least have a looksie:

$$\frac{d}{dx} \arcsin x = \frac{1}{\sqrt{1 - x^2}}$$

$$\frac{d}{dx} \arccos x = -\frac{1}{\sqrt{1 - x^2}}$$

$$\frac{d}{dx} \arctan x = \frac{1}{1 + x^2}$$

$$\frac{d}{dx} \operatorname{arccot} x = -\frac{1}{1 + x^2}$$

$$\frac{d}{dx}\operatorname{arcsec}x = \frac{1}{x\sqrt{x^2-1}}$$

$$\frac{d}{dx}\operatorname{arccsc}x = -\frac{1}{x\sqrt{x^2-1}}$$

Notice that derivatives of the three "co" functions are just negations of the three other functions, so your work is cut in half.

The Sum Rule

In textbooks, the Sum Rule is often phrased: The derivative of the sum of functions equals the sum of the derivatives of those functions:

$$\frac{d}{dx}[f(x)+g(x)] = \frac{d}{dx}f(x) + \frac{d}{dx}g(x)$$

Simply put, the Sum Rule tells you that differentiating long expressions term by term is okay. For example, suppose that you want to evaluate the following:

$$\frac{d}{dx}(\sin x + x^4 - \ln x)$$

The expression that you're differentiating has three terms, so by the Sum Rule, you can break this into three separate derivatives and solve them separately:

$$= \frac{d}{dx}\sin x + \frac{d}{dx}x^4 - \frac{d}{dx}\ln x$$

$$= \cos x + 4x^3 - \frac{1}{x}$$

Note that the Sum Rule also applies to expressions of more than two terms. It also applies regardless of whether the term is positive or negative. Some books call this variation the Difference Rule, but you get the idea.

The Constant Multiple Rule

A typical textbook gives you this sort of definition for the Constant Multiple Rule: The derivative of a constant multiplied by a function equals the product of that constant and the derivative of that function:

$$\frac{d}{dx}nf(x) = n\frac{d}{dx}f(x)$$

In plain English, this rule tells you that moving a constant outside of a derivative before you differentiate is okay. For example:

$$\frac{d}{dx}5\tan x$$

To solve this, move the 5 outside the derivative, and then differentiate:

$$= 5\frac{d}{dx}\tan x$$

$$= 5\sec^2 x$$

The Power Rule

The Power Rule tells you that to find the derivative of x raised to any power, bring down the exponent as the coefficient of x, and then subtract 1 from the exponent and use this as your *new* exponent. Here's the general form:

$$\frac{d}{dx}\, x^n = nx^{n-1}$$

Here are a few examples:

$$\frac{d}{dx}\, x^2 = 2x$$

$$\frac{d}{dx}\, x^3 = 3x^2$$

$$\frac{d}{dx}\, x^{10} = 10x^9$$

When the function that you're differentiating already has a coefficient, multiply the exponent by this coefficient. For example:

$$\frac{d}{dx}\, 2x^4 = 8x^3$$

$$\frac{d}{dx}\, 7x^6 = 42x^5$$

$$\frac{d}{dx}\, 4x^{100} = 400x^{99}$$

The Power Rule also extends to negative exponents, which allows you to differentiate many fractions. For example:

$$\frac{d}{dx}\, \frac{1}{x^5}$$

$$= \frac{d}{dx}\, x^{-5}$$

$$= -5x^{-6}$$

$$= -\frac{5}{x^6}$$

It also extends to fractional exponents, which allows you to differentiate square roots and other roots:

$$\left(\frac{d}{dx}\right) x^{\frac{1}{3}} = \frac{1}{3}\, x^{-\frac{2}{3}}$$

The Product Rule

The derivative of the product of two functions $f(x)$ and $g(x)$ is equal to the derivative of $f(x)$ multiplied by $g(x)$ plus the derivative of $g(x)$ multiplied by $f(x)$. That is:

$$\frac{d}{dx}\left[f(x) \cdot g(x)\right]$$

$$= f'(x) \cdot g(x) + g'(x) \cdot f(x)$$

Practice saying the Product Rule like this: "The derivative of the first function times the second *plus* the derivative of the second times the first." This encapsulates the Product Rule and sets you up to remember the Quotient Rule (see the next section).

For example, suppose that you want to differentiate $e^x \sin x$. Start by breaking the problem out as follows:

$$\frac{d}{dx} e^x \sin x = \left(\frac{d}{dx} e^x\right)\sin x + \left(\frac{d}{dx} \sin x\right)e^x$$

Now, you can evaluate both derivatives, which I underline, without much confusion:

$$= e^x \cdot \sin x + \cos x \cdot e^x$$

You can clean this up a bit as follows:

$$= e^x\left(\sin x + \cos x\right)$$

The Quotient Rule

$$\frac{d}{dx}\left(\frac{f(x)}{g(x)}\right) = \frac{f'(x) \cdot g(x) - g'(x) \cdot f(x)}{g(x)^2}$$

Practice saying the Quotient Rule like this: "The derivative of the top function times the bottom *minus* the derivative of the bottom times the top, over the bottom squared." This is similar enough to the Product Rule that you can remember it.

For example, suppose that you want to differentiate the following:

$$\frac{d}{dx}\left(\frac{x^4}{\tan x}\right)$$

As you do with the Product Rule example, start by breaking the problem out as follows:

$$= \frac{\left(\frac{d}{dx}x^4\right) \cdot \tan x - \left(\frac{d}{dx}\tan x\right) \cdot x^4}{\tan^2 x}$$

Now, evaluate the two derivatives:

$$= \frac{4x^3 \cdot \tan x - \sec^2 x \cdot x^4}{\tan^2 x}$$

This answer is fine, but you can clean it up by using some algebra plus the five basic trig identities from earlier in this chapter. (Don't worry too much about these steps unless your professor is particularly unforgiving.)

$$= \frac{4x^3 \tan x}{\tan^2 x} - (x^4 \sec^2 x \cot^2 x)$$

$$= \frac{4x^3}{\tan x} - x^4\left(\frac{1}{\cos^2 x}\right)\left(\frac{\cos^2 x}{\sin^2 x}\right)$$

$$= 4x^3 \cot x - x^4 \csc^2 x$$

$$= x^3 (4 \cot x - x \csc^2 x)$$

The Chain Rule

I'm aware that the Chain Rule is considered a major sticking point in Calculus I, so I take a little time to review it. (By the way, contrary to popular belief, the Chain Rule isn't "If you don't follow the *rules* in your Calculus class, the teacher gets to place you in *chains*." Such teaching methods are now considered questionable and have been out of use in the classroom since at least the 1970s.)

The Chain Rule allows you to differentiate nested functions — that is, functions within functions. It places no limit on how deeply nested these functions are. In this section, I show you an easy way to think about nested functions, and then I show you how to apply the Chain Rule simply.

Evaluating functions from the inside out

When you're evaluating a nested function, you begin with the *inner* function and move *outward*. For example:

$$f(x) = e^{2x}$$

In this case, $2x$ is the inner function. To see why, suppose that you want to evaluate $f(x)$ for a given value of x. To keep things simple, say that $x = 0$. After plugging in 0 for x, your first step is to evaluate the inner function, which I underline:

Step 1: $e^{2(0)} = e^0$

Your next step is to evaluate the outer function:

Step 2: $\underline{e^0} = 1$

The terms *inner function* and *outer function* are determined by the order in which the functions get evaluated. This is true no matter how deeply nested these functions are. For example:

$$g(x) = \left(\ln \sqrt{e^{3x-6}} \right)^3$$

Suppose that you want to evaluate $g(x)$. To keep the numbers simple, this time let $x = 2$. After plugging in 2 for x, here's the order of evaluation from the inner function to the outer:

Step 1: $\left(\ln \sqrt{e^{3(2)-6}} \right)^3 = \left(\ln \sqrt{e^0} \right)^3$

Step 2: $\left(\ln \sqrt{e^0} \right)^3 = \left(\ln \sqrt{1} \right)^3$

Step 3: $\left(\ln \sqrt{1} \right)^3 = (\ln 1)^3$

Step 4: $(\ln 1)^3 = 0^3$

Step 5: $0^3 = 0$

The process of evaluation clearly lays out the five nested functions of $g(x)$ from inner to outer.

Differentiating functions from the outside in

In contrast to evaluation, differentiating a function by using the Chain Rule forces you to begin with the *outer* function and move *inward*.

Here's the basic Chain Rule the way that you find it in textbooks:

$$\frac{d}{dx} f(g(x)) = f'(g(x)) \cdot g'(x)$$

To differentiate nested functions by using the Chain Rule, write down the derivative of the outer function, copying everything inside it, and multiply this result by the derivative of the next function inward.

This explanation may seem a bit confusing, but it's a lot easier when you know how to find the outer function, which I explain in the previous section, "Evaluating functions from the inside out." A couple of examples should help.

For example, suppose that you want to differentiate the nested function $\sin 2x$. The outer function is the sine portion, so this is where you start:

$$\frac{d}{dx} \sin 2x = \cos 2x \cdot \frac{d}{dx} 2x$$

To finish, you still need to differentiate the underlined portion, $2x$:

$$= \cos 2x \cdot 2$$

Rearranging this solution to make it more presentable gives you your final answer:

$$= 2 \cos 2x$$

When you differentiate more than two nested functions, the Chain Rule really lives up to its name: As you break down the problem step by step, you string out a *chain* of multiplied expressions.

For example, suppose that you want to differentiate $\sin^3 e^x$. Remember from the earlier section, "Noting trig notation," that the notation $\sin^3 e^x$ really means $(\sin e^x)^3$. This rearrangement makes clear that the outer function is the power of 3, so begin differentiating with this function:

$$\frac{d}{dx} (\sin e^x)^3 = 3(\sin e^x)^2 \cdot \frac{d}{dx} (\sin e^x)$$

Now, you have a smaller function to differentiate, which I underline:

$$= 3(\sin e^x)^2 \cdot \cos e^x \cdot \frac{d}{dx} e^x$$

Only one more derivative to go:

$$= 3(\sin e^x)^2 \cdot \cos e^x \cdot e^x$$

Again, rearranging your answer is customary:

$$= 3e^x \cos e^x \sin^2 e^x$$

Finding Limits by Using L'Hospital's Rule

L'Hospital's Rule is all about limits and derivatives, so it fits better with Calculus I than Calculus II. But some colleges save this topic for Calculus II.

So, even though I'm addressing this as a review topic, fear not: Here, I give you the full story of L'Hospital's Rule, starting with how to pronounce L'Hospital (low-pee-tahl).

L'Hospital's Rule provides a method for evaluating certain *indeterminate forms* of limits. First, I show you what an indeterminate form of a limit looks like, with a list of all common indeterminate forms. Next, I show you how to use L'Hospital's Rule to evaluate some of these forms. And finally, I show you how to work with the other indeterminate forms so that you can evaluate them.

Understanding determinate and indeterminate forms of limits

As you discover earlier in this chapter, in "Knowing your limits," you can evaluate many limits by simply replacing the limit variable with the number that it approaches. In some cases, this replacement results in a number, so this number is the value of the limit that you're seeking. In other cases, this replacement gives you an infinite value (either $+\infty$ or $-\infty$), so the limit does not exist (DNE).

Table 2-6 shows a list of some functions that often cause confusion.

Table 2-6		Limits of Some Common Functions		
Case	*f(x) =*	*g(x) =*	*Function*	*Limit*
#1	0	∞	$\dfrac{f(x)}{g(x)}$	0
#2	0	∞	$f(x)^{g(x)}$	0
#3	$C \neq 0$	0	$\dfrac{f(x)}{g(x)}$	DNE
#4	$\pm\infty$	0	$\dfrac{f(x)}{g(x)}$	DNE

To understand how to think about these four cases, remember that a limit describes the behavior of a function very close to, but not exactly at, a value of x.

In the first and second cases, $f(x)$ gets very close to 0 and $g(x)$ explodes to infinity, so both $\dfrac{f(x)}{g(x)}$ and $f(x)^{g(x)}$ approach 0. In the third case, $f(x)$ is a constant c other than 0 and $g(x)$ approaches 0, so the fraction $\dfrac{f(x)}{g(x)}$ explodes to infinity. And in the fourth case, $f(x)$ explodes to infinity and $g(x)$ approaches 0, so the fraction $\dfrac{f(x)}{g(x)}$ explodes to infinity.

In each of these cases, you have the answer you're looking for — that is, you know whether the limit exists and, if so, its value — so these are all called *determinate forms* of a limit.

In contrast, however, sometimes when you try to evaluate a limit by replacement, the result is an *indeterminate form* of a limit. Table 2-7 includes two common indeterminate forms.

Table 2-7		Two Indeterminate Forms of Limits		
Case	*f(x) =*	*g(x) =*	*Function*	*Limit*
#1	0	0	$\dfrac{f(x)}{g(x)}$	Indeterminate
#2	$\pm\infty$	$\pm\infty$	$\dfrac{f(x)}{g(x)}$	Indeterminate

In these cases, the limit becomes a race between the numerator and denominator of the fractional function. For example, think about the second example in the chart. If $f(x)$ crawls toward ∞ while $g(x)$ zooms there, the fraction becomes bottom heavy and the limit is 0.

But if $f(x)$ zooms to ∞ while $g(x)$ crawls there, the fraction becomes top heavy and the limit is ∞ — that is, DNE. And if both functions move toward 0 proportionally, this proportion becomes the value of the limit.

When attempting to evaluate a limit by replacement saddles you with either of these forms, you need to do more work. Applying L'Hospital's Rule is the most reliable way to get the answer that you're looking for.

Introducing L'Hospital's Rule

Suppose that you're attempting to evaluate the limit of a function of the form $\dfrac{f(x)}{g(x)}$. When replacing the limit variable with the number that it approaches results in either $\dfrac{0}{0}$ or $\dfrac{\pm\infty}{\pm\infty}$, L'Hospital's Rule tells you that the following equation holds true:

$$\lim_{x \to c} \frac{f(x)}{g(x)} = \lim_{x \to c} \frac{f'(x)}{g'(x)}$$

Note that c can be any real number as well as ∞ or $-\infty$.

As an example, suppose that you want to evaluate the following limit:

$$\lim_{x \to 0} \frac{x^3}{\sin x}$$

Replacing x with 0 in the function leads to the following result:

$$\frac{0^3}{\sin 0} = \frac{0}{0}$$

This is one of the two indeterminate forms that L'Hospital's Rule applies to, so you can draw the following conclusion:

$$\lim_{x \to 0} \frac{x^3}{\sin x} = \lim_{x \to 0} \frac{(x^3)'}{(\sin x)'}$$

Next, evaluate the two derivatives:

$$= \lim_{x \to 0} \frac{3x^2}{\cos x}$$

Now, use this new function to try another replacement of x with 0 and see what happens:

$$\frac{3(0^2)}{\cos 0} = \frac{0}{1}$$

This time, the result is a determinate form, so you can evaluate the original limit as follows:

$$\lim_{x \to 0} \frac{x^3}{\sin x} = 0$$

In some cases, you may need to apply L'Hospital's Rule more than once to get an answer. For example:

$$\lim_{x \to \infty} \frac{e^x}{x^5}$$

Replacement of x with ∞ results in the indeterminate form $\frac{\infty}{\infty}$, so you can use L'Hospital's Rule:

$$\lim_{x \to \infty} \frac{e^x}{e^5} = \lim_{x \to \infty} \frac{e^x}{5x^4}$$

In this case, the new function gives you the same indeterminate form, so use L'Hospital's Rule again:

$$= \lim_{x \to \infty} \frac{e^x}{20x^3}$$

The same problem arises, but again you can use L'Hospital's Rule. You can probably see where this example is going, so I fast forward to the end:

$$= \lim_{x \to \infty} \frac{e^x}{60x^2}$$

$$= \lim_{x \to \infty} \frac{e^x}{120x}$$

$$= \lim_{x \to \infty} \frac{e^x}{120}$$

When you apply L'Hospital's Rule repeatedly to a problem, make sure that every step along the way results in one of the two indeterminate forms that the rule applies to.

At last! The process finally yields a function with a determinate form:

$$\frac{e^\infty}{120} = \frac{\infty}{120} = \infty$$

Therefore, the limit does not exist.

Alternative indeterminate forms

L'Hospital's Rule applies only to the two indeterminate forms $\frac{0}{0}$ and $\frac{\pm\infty}{\pm\infty}$.

But limits can result in a variety of other indeterminate forms for which L'Hospital's Rule doesn't hold. Table 2-8 is a list of the indeterminate forms that you're most likely to see.

Table 2-8	Five Cases of Indeterminate Forms Where You Can't Apply L'Hospital's Rule Directly			
Case	*f(x) =*	*g(x) =*	*Function*	*Form*
#1	0	∞	$f(x) \cdot g(x)$	Indeterminate
#2	∞	∞	$f(x) - g(x)$	Indeterminate
#3	0	0	$f(x)^{g(x)}$	Indeterminate
#4	∞	0	$f(x)^{g(x)}$	Indeterminate
#5	1	∞	$f(x)^{g(x)}$	Indeterminate

Because L'Hospital's Rule doesn't hold for these indeterminate forms, applying the rule directly gives you the wrong answer.

These indeterminate forms require special attention. In this section, I show you how to rewrite these functions so that you can then apply L'Hospital's Rule.

Case #1: $0 \cdot \infty$

When $f(x) = 0$ and $g(x) = \infty$, the limit of $f(x) \cdot g(x)$ is the indeterminate form $0 \cdot \infty$, which doesn't allow you to use L'Hospital's Rule. To evaluate this limit, rewrite this function as follows:

$$f(x) \cdot g(x) = \frac{f(x)}{\dfrac{1}{g(x)}}$$

The limit of this new function is the indeterminate form $\frac{0}{0}$, which allows you to use L'Hospital's Rule. For example, suppose that you want to evaluate the following limit:

$$\lim_{x \to 0^+} x \cot x$$

Replacing x with 0 gives you the indeterminate form $0 \cdot \infty$, so rewrite the limit as follows:

$$= \lim_{x \to 0^+} \frac{x}{\left(\dfrac{1}{\cot x}\right)}$$

This can be simplified a little by using the inverse trig identity for cot x:

$$= \lim_{x \to 0^+} \frac{x}{\tan x}$$

Now, replacing x with 0 gives you the indeterminate for $\frac{0}{0}$, so you can apply L'Hospital's Rule.

$$= \lim_{x \to 0^+} \frac{(x)'}{(\tan x)'}$$

$$= \lim_{x \to 0^+} \frac{1}{\sec^2 x}$$

At this point, you can evaluate the limit directly by replacing x with 0:

$$= \frac{1}{1} = 1$$

Therefore, the limit evaluates to 1.

Case #2: $\infty - \infty$

When $f(x) = \infty$ and $g(x) = \infty$, the limit of $f(x) - g(x)$ is the indeterminate form $\infty - \infty$, which doesn't allow you to use L'Hospital's Rule. To evaluate this limit, try to find a common denominator that turns the subtraction into a fraction. For example:

$$\lim_{x \to 0^+} \cot x - \csc x$$

In this case, replacing x with 0 gives you the indeterminate form $\infty - \infty$. A little tweaking with the Basic Five trig identities (see "Identifying some important trig identities" earlier in this chapter) does the trick:

$$= \lim_{x \to 0^+} \frac{\cos x}{\sin x} - \frac{1}{\sin x}$$

$$= \lim_{x \to 0^+} \frac{\cos x - 1}{\sin x}$$

Now, replacing x with 0 gives you the indeterminate form $\frac{0}{0}$, so you can use L'Hospital's Rule:

$$= \lim_{x \to 0^+} \frac{(\cos x - 1)'}{(\sin x)'}$$

$$= \lim_{x \to 0^+} \frac{-\sin x}{\cos x}$$

At last, you can evaluate the limit by directly replacing x with 0.

$$= \frac{0}{1} = 0$$

Therefore, the limit evaluates to 0.

Cases #3, #4, and #5: 0^0, ∞^0, and 1^∞

In the following three cases, the limit of $f(x)^{g(x)}$ is an indeterminate form that doesn't allow you to use L'Hospital's Rule:

- ✔ When $f(x) = 0$ and $g(x) = 0$
- ✔ When $f(x) = \infty$ and $g(x) = 0$
- ✔ When $f(x) = 1$ and $g(x) = \infty$

This indeterminate form 1^∞ is easy to forget because it seems weird. After all, $1^x = 1$ for every real number, so why should 1^∞ be any different? In this case, infinity plays one of its many tricks on mathematics. You can find out more about some of these tricks in Chapter 16.

For example, suppose that you want to evaluate the following limit:

$$\lim_{x \to 0} x^x$$

As it stands, this limit is of the indeterminate form 0^0.

Fortunately, I can show you a trick to handle these three cases. As with so many things mathematical, mere mortals such as you and me probably wouldn't discover this trick, short of being washed up on a desert island with nothing to do but solve math problems and eat coconuts. However, somebody did the hard work already. Remembering this following recipe is a small price to pay:

1. **Set the limit equal to *y*.**

   $$y = \lim_{x \to 0} x^x$$

2. **Take the natural log of both sides, and then do some *log rolling*:**

   $$\ln y = \ln \lim_{x \to 0} x^x$$

   Here are the two log rolling steps:

   - First, *roll* the log inside the limit:

     $$= \lim_{x \to 0} \ln x^x$$

     This step is valid because the limit of a log equals the log of a limit (I know, those words veritably *roll* off the tongue).

   - Next, *roll* the exponent over the log:

     $$= \lim_{x \to 0} x \ln x$$

     This step is also valid, as I show you earlier in this chapter when I discuss the log function in "Graphing common functions."

3. **Evaluate this limit as I show you in "Case #1: $0 \cdot \infty$."**

   Begin by changing the limit to a determinate form:

   $$= \lim_{x \to 0} \frac{\ln x}{\frac{1}{x}}$$

   At last, you can apply L'Hospital's Rule:

   $$= \lim_{x \to 0} \frac{(\ln x)'}{\left(\frac{1}{x}\right)'}$$

   $$= \lim_{x \to 0} \frac{\frac{1}{x}}{-\frac{1}{x^2}}$$

   Now, evaluating the limit isn't too bad:

   $$= \lim_{x \to 0} -\frac{x^2}{x}$$

   $$= \lim_{x \to 0} -x = 0$$

   Wait! Remember that way back in Step 2 you set this limit equal to ln *y*. So you have one more step!

4. Solve for *y*.

$$\ln y = 0$$

$$y = 1$$

Yes, this is your final answer, so $\lim_{x \to 0} x^x = 1$.

This recipe works with all three indeterminate forms that I talk about at the beginning of this section. Just make sure that you keep tweaking the limit until you have one of the two forms that are compatible with L'Hospital's Rule.

Chapter 3

From Definite to Indefinite: The Indefinite Integral

The first step to solving an area problem — that is, finding the area of a complex or unusual shape on the graph — is expressing it as a definite integral. In turn, you can evaluate a definite integral by using a formula based on the limit of a Riemann sum (as I show you in Chapter 1).

In this chapter, you get down to business calculating definite integrals. First, I show you a variety of different ways to estimate area. All these methods lead to a better understanding of the Riemann sum formula for the definite integral. Next, you use this formula to find exact areas. This rather hairy method of calculating definite integrals prompts a search for a better way.

This better way is the indefinite integral. I show you how the indefinite integral provides a much simpler way to calculate area. Furthermore, you find a surprising link between differentiation (which is the focus of Calculus I) and integration. This link, called the Fundamental Theorem of Calculus, shows that the indefinite integral is really an *anti-derivative* (the inverse of the derivative).

To finish up, I show you how using an indefinite integral to evaluate a definite integral results in signed area. I also clarify the differences between definite and indefinite integrals so that you never get them confused. By the end of this chapter, you're ready for Part II, which focuses on an abundance of methods for calculating the indefinite integral.

Approximate Integration

Finding the *exact* area under a curve — that is, solving an area problem (see Chapter 1) — is one of the main reasons that integration was invented. But you can approximate area by using a variety of methods. Approximating area is a good first step toward understanding how integration works.

In this section, I show you five different methods for approximating the solution to an area problem. Generally speaking, I introduce these methods in the order of increasing difficulty and effectiveness. The first three involve manipulating rectangles.

- ✔ The first two methods — left and right rectangles — are the easiest to use, but they usually give you the greatest margin of error.

- ✔ The Midpoint Rule (slicing rectangles) is a little more difficult, but it usually gives you a slightly better estimate.

- ✔ The Trapezoid Rule requires more computation, but it gives an even better estimate.

- ✔ Simpson's Rule is the most difficult to grasp, but it gives the best approximation and, in some cases, provides you with an exact measurement of area.

Three ways to approximate area with rectangles

Slicing an irregular shape into rectangles is the most common approach to approximating its area (see Chapter 1 for more details on this approach). In this section, I show you three different techniques for approximating area with rectangles.

Using left rectangles

You can use left rectangles to approximate the solution to an area problem (see Chapter 1). For example, suppose that you want to approximate the shaded area in Figure 3-1 by using four left rectangles.

To draw these four rectangles, start by dropping a vertical line from the function to the x-axis at the *left-hand* limit of integration — that is, $x = 0$. Then drop three more vertical lines from the function to the x-axis at $x = 2$, 4, and 6. Next, at the four points where these lines cross the function, draw horizontal lines *from left to right* to make the top edges of the four rectangles. The left and top edges define the size and shape of each left rectangle.

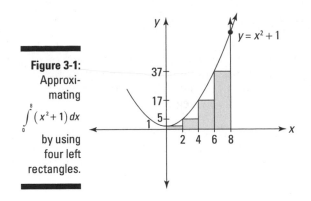

Figure 3-1:
Approxi-
mating
$\int_{0}^{8}(x^2+1)\,dx$
by using
four left
rectangles.

To measure the areas of these four rectangles, you need the width and height of each. The width of each rectangle is obviously 2. The height and area of each is determined by the value of the function at its *left* edge, as shown in Table 3-1.

Table 3-1	Approximating Area by Using Left Rectangles		
Rectangle	*Width*	*Height*	*Area*
#1	2	$0^2+1=1$	2
#2	2	$2^2+1=5$	10
#3	2	$4^2+1=17$	34
#4	2	$6^2+1=37$	74

To approximate the shaded area, add up the areas of these four rectangles:

$$\int_{0}^{8}(x^2+1)\,dx \approx 2+10+34+74=120$$

Using right rectangles

Using right rectangles to approximate the solution to an area problem is virtually the same as using left rectangles. For example, suppose that you want to use six right rectangles to approximate the shaded area in Figure 3-2.

To draw these rectangles, start by dropping a vertical line from the function to the *x*-axis at the *right-hand* limit of integration — that is, $x = 3$. Next, drop five more vertical lines from the function to the *x*-axis at $x = 0.5$, 1, 1.5, 2, and 2.5. Then, at the six points where these lines cross the function, draw horizontal lines *from right to left* to make the top edges of the six rectangles. The right and top edges define the size and shape of each left rectangle.

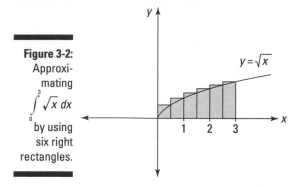

Figure 3-2:
Approximating
$$\int_0^3 \sqrt{x}\, dx$$
by using
six right
rectangles.

To measure the areas of these six rectangles, you need the width and height of each. Each rectangle's width is 0.5. Its height and area are determined by the value of the function at its *right* edge, as shown in Table 3-2.

Table 3-2	Approximating Area by Using Right Rectangles		
Rectangle	*Width*	*Height*	*Area*
#1	0.5	$\sqrt{0.5} \approx 0.707$	0.354
#2	0.5	$\sqrt{1} = 1$	0.5
#3	0.5	$\sqrt{1.5} \approx 1.225$	0.613
#4	0.5	$\sqrt{2} \approx 1.414$	0.707
#5	0.5	$\sqrt{2.5} \approx 1.581$	0.791
#6	0.5	$\sqrt{3} \approx 1.732$	0.866

To approximate the shaded area, add up the areas of these six rectangles:

$$\int_0^3 \sqrt{x}\, dx \approx 0.354 + 0.5 + 0.613 + 0.707 + 0.791 + 0.866 = 3.831$$

Finding a middle ground: The Midpoint Rule

Both left and right rectangles give you a decent approximation of area. So, it stands to reason that slicing an area vertically and measuring the height of each rectangle from the *midpoint* of each slice might give you a slightly better approximation of area.

For example, suppose that you want to use midpoint rectangles to approximate the shaded area in Figure 3-3.

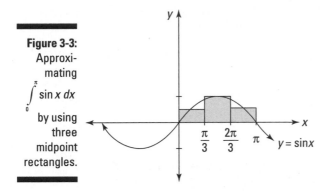

To draw these three rectangles, start by drawing vertical lines that intersect both the function and the x-axis at $x = 0$, $\frac{\pi}{3}$, $\frac{2\pi}{3}$, and π. Next, find where the midpoints of these three regions — that is, $\frac{\pi}{6}$, $\frac{\pi}{2}$, and $\frac{5\pi}{6}$ — intersect the function. Now, draw horizontal lines through these three points to make the tops of the three rectangles.

To measure these three rectangles, you need the width and height of each to compute the area. The width of each rectangle is $\frac{\pi}{3}$, and the height is given in Table 3-3.

Table 3-3   Approximating Area by Using the Midpoint Rule

Rectangle	Width	Height	Area
#1	$\frac{\pi}{3}$	$\sin \frac{\pi}{6} = \frac{1}{2}$	$\frac{\pi}{6}$
#2	$\frac{\pi}{3}$	$\sin \frac{\pi}{2} = 1$	$\frac{\pi}{3}$
#3	$\frac{\pi}{3}$	$\sin \frac{\pi}{6} = \frac{1}{2}$	$\frac{\pi}{6}$

To approximate the shaded area, add up the areas of these three rectangles:

$$\int_0^\pi \sin x \, dx \approx \frac{\pi}{6} + \frac{\pi}{3} + \frac{\pi}{6} = \frac{2\pi}{3} \approx 2.0944$$

The slack factor

The formula for the definite integral is based on Riemann sums (see Chapter 1). This formula allows you to add up the area of infinitely many infinitely thin rectangular slices to find the exact solution to an area problem.

And here's the strange part: Within certain parameters, the Riemann sum formula doesn't care *how* you do the slicing. All three slicing methods that I discuss in the previous section work equally well. That is, although each method yields a different *approximate* area for a given *finite* number of slices, all these differences are smoothed over when the limit is applied. In other words, all three methods work to provide you the *exact* area for *infinitely many* slices.

I call this feature of measuring rectangles the slack factor. Understanding the slack factor helps you understand why using rectangles drawn at the left endpoint, right endpoint, or midpoint all lead to the same *exact* value of an area: As you measure progressively thinner slices, the slack factor never increases and tends to decrease. As the number of slices approaches ∞, the width of each slice approaches 0, so the slack factor also approaches 0.

Figure 3-4 shows the range of this slack in choosing a rectangle. In this example, to find the area under $f(x)$, you need to measure a rectangle inside the given slice. The height of this rectangle must be inclusively between p and q, the local maximum and minimum of $f(x)$. Within these parameters, however, you can measure *any* rectangle.

Figure 3-4:
For each slice you're measuring, you can use *any* rectangle that passes through the function at one point or more.

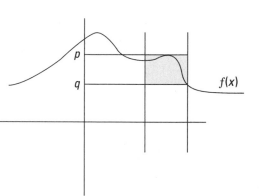

Two more ways to approximate area

Although slicing a region into rectangles is the simplest way to approximate its area, rectangles aren't the only shape that you can use. For finding many areas, other shapes can yield a better approximation in fewer slices.

In this section, I show you two common alternatives to rectangular slicing: the Trapezoid Rule (which, not surprisingly, uses trapezoids) and Simpson's Rule (which uses rectangles topped with parabolas).

Feeling trapped? The Trapezoid Rule

In case you feel restricted — dare I say *boxed in?* — by estimating areas with only rectangles, you can get an even closer approximation by drawing trapezoids instead of rectangles.

For example, suppose that you want to use six trapezoids to estimate this area:

$$\int_{-3}^{3} 9 - x^2 \, dx$$

You can probably tell just by looking at the graph in Figure 3-5 that using trapezoids gives you a closer approximation than rectangles. In fact, the area of a trapezoid drawn on any slice of a function will be the average of the areas of the left and right rectangles drawn on that slice.

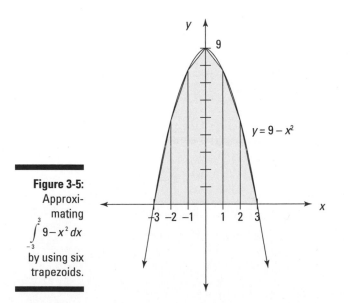

Figure 3-5:
Approximating $\int_{-3}^{3} 9 - x^2 \, dx$ by using six trapezoids.

To draw these six trapezoids, first plot points along the function at $x = -3, -2,$ $-1, 0, 1, 2,$ and 3. Next, connect adjacent points to make the top edges of the trapezoids. Finally, draw vertical lines through these points.

Two of the six "trapezoids" are actually triangles. This fact doesn't affect the calculation; just think of each triangle as a trapezoid with one height equal to zero.

To find the area of these six trapezoids, use the formula for the area of a trapezoid that you know from geometry: $\dfrac{w(b_1 + b_2)}{2}$. In this case, however, the two bases — that is, the parallel sides of the trapezoid — are the heights on the left and right sides. As always, the width is easy to calculate — in this case, it's 1. Table 3-4 shows the rest of the information for calculating the area of each trapezoid.

Table 3-4		Approximating Area by Using Trapezoids		
Trapezoid	*Width*	*Left Height*	*Right Height*	*Area*
#1	1	$9 - (-3)^2 = 0$	$9 - (-2)^2 = 5$	$\dfrac{1(0+5)}{2} = 2.5$
#2	1	$9 - (-2)^2 = 5$	$9 - (-1)^2 = 8$	$\dfrac{1(5+8)}{2} = 6.5$
#3	1	$9 - (-1)^2 = 8$	$9 - (0)^2 = 9$	$\dfrac{1(8+9)}{2} = 8.5$
#4	1	$9 - (0)^2 = 9$	$9 - (1)^2 = 8$	$\dfrac{1(9+8)}{2} = 8.5$
#5	1	$9 - (1)^2 = 8$	$9 - (2)^2 = 5$	$\dfrac{1(8+5)}{2} = 6.5$
#6	1	$9 - (2)^2 = 5$	$9 - (3)^2 = 0$	$\dfrac{1(5+0)}{2} = 2.5$

To approximate the shaded area, find the sum of the six areas of the trapezoids:

$$\int_{-3}^{3} 9 - x^2 \, dx \approx 2.5 + 6.5 + 8.5 + 8.5 + 6.5 + 2.5 = 35$$

Don't have a cow! Simpson's Rule

You may recall from geometry that you can draw exactly one circle through any three nonlinear points. You may not recall, however, that the same is true of parabolas: Just three nonlinear points determine a parabola.

Simpson's Rule relies on this geometric theorem. When using Simpson's Rule, you use left and right endpoints as well as midpoints as these three points for each slice.

1. **Begin slicing the area that you want to approximate into strips that intersect the function.**

2. **Mark the left endpoint, midpoint, and right endpoint of each strip.**

3. **Top each strip with the section of the parabola that passes through these three points.**

4. **Add up the areas of these parabola-topped strips.**

At first glance, Simpson's Rule seems a bit circular: You're trying to approximate the area under a curve, but this method forces you to measure the area inside a region that includes a curve. Fortunately, Thomas Simpson, who invented this rule, is way ahead on this one. His method allows you to measure these strangely shaped regions without too much difficulty.

Without further ado, here's Simpson's Rule:

Given that n is an even number,

$$\int f(x)\,dx \infty f(x)\,dx$$

$$\approx \frac{b-a}{3n} \left[f(x_0) + 4f(x_1) + 2f(x_2) + 4f(x_0) + \ldots + 4f(x_3) + 2f(x_0) + 4f(x_0) + f(x_0) \right]$$

What does it all mean? As with every approximation method you've encountered, the key to Simpson's Rule is measuring the width and height of each of these regions (with some adjustments):

✔ The width is represented by $\frac{b-a}{n}$ — but Simpson's Rule adjusts this value to $\frac{b-a}{3n}$.

✔ The heights are represented by $f(x)$ taken at various values of x — but Simpson's Rule multiplies some of these by a coefficient of either 4 or 2. (By the way, these choices of coefficients are based on the known result of the area under a parabola — not just picked out of the air!)

The best way to show you how this rule works is with an example. Suppose that you want to use Simpson's Rule to approximate the following:

$$\int_1^5 \frac{1}{x}\,dx$$

First, divide the area that you want to approximate into an *even* number of regions — say, eight — by drawing nine vertical lines from $x = 1$ to $x = 5$. Now top these regions off with parabolas as I show you in Figure 3-6.

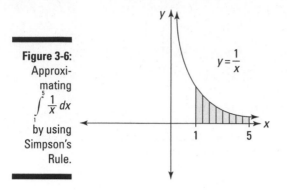

Figure 3-6:
Approxi-
mating
$\int\limits_{1}^{5} \frac{1}{x} dx$
by using
Simpson's
Rule.

The width of each region is 0.5, so adjust this by dividing by 3:

$$\frac{b-a}{3n} = \frac{0.5}{3} \approx 0.167$$

Moving on to the heights, find $f(x)$ when $x = 1, 1.5, 2, \dots, 4.5,$ and 5 (see the second column of Table 3-5). Adjust all these values except the first and the last by multiplying by 4 or 2, alternately.

Table 3-5	Approximating Area by Using Simpson's Rule		
n	*f(x$_n$)*	*Coefficient*	*Total*
0	$f(1) = 1$	1	$f(1) = 1$
1	$f(1.5) = 0.667$	4	$4f(1.5) = 2.668$
2	$f(2) = 0.5$	2	$2f(2) = 1$
3	$f(2.5) = 0.4$	4	$4f(2.5) = 1.6$
4	$f(3) = 0.333$	2	$2f(3) = 0.666$
5	$f(3.5) = 0.290$	4	$4f(3.5) = 1.160$
6	$f(4) = 0.25$	2	$2f(4) = 0.5$
7	$f(4.5) = 0.222$	4	$4f(4.5) = 0.888$
8	$f(5) = 0.2$	1	$f(5) = 0.2$

Now, apply Simpson's Rule as follows:

$$\int\limits_{1}^{5} \frac{1}{x}\, dx$$

$$\approx 0.167 \,(1 + 2.668 + 1 + 1.6 + 0.666 + 1.16 + 0.5 + 0.888 + 0.2)$$

$$= 0.167 \,(9.682) \approx 1.617$$

So Simpson's Rule approximates the area of the shaded region in Figure 3-6 as 1.617. (By the way, the actual area to three decimal places is about 1.609 — so Simpson's Rule provides a pretty good estimate.)

In fact, Simpson's Rule often provides an even better estimate than this example leads you to believe, because a lot of inaccuracy arises from rounding off decimals. In this case, when you perform the calculations with enough precision, Simpson's Rule provides the correct area to three decimal places!

Knowing Sum-Thing about Summation Formulas

In Chapter 1, I introduce you to the Riemann sum formula for the definite integral. This formula includes a summation using sigma notation (Σ). (Please flip to Chapter 2 if you need a refresher on this topic.)

In practice, evaluating a summation can be a little tricky. Fortunately, three important summation formulas exist to help you. In this section, I introduce you to these formulas and show you how to use them. In the next section, I show you how and when to apply them when you're using the Riemann sum formula to solve an area problem.

The summation formula for counting numbers

The summation formula for counting numbers gives you an easy way to find the sum $1 + 2 + 3 + \dots + n$ for any value of n:

$$\sum_{i=1}^{n} i = \frac{n(n+1)}{2}$$

To see how this formula works, suppose that $n = 9$:

$$\sum_{i=1}^{9} i = 1 + 2 + 3 + 4 + 5 + 6 + 7 + 8 + 9 = 45$$

The summation formula for counting numbers also produces this result:

$$\frac{n(n+1)}{2} = \frac{9(10)}{2} = 45$$

According to a popular story, mathematician Karl Friedrich Gauss discovered this formula as a schoolboy, when his teacher gave the class the boring task of adding up all the counting numbers from 1 to 100 so that he (the teacher) could nap at his desk. Within minutes, Gauss arrived at the correct answer, 5,050, disturbing his teacher's snooze time and making mathematical history.

The summation formula for square numbers

The summation formula for square numbers gives you a quick way to add up $1 + 4 + 9 + \ldots + n^2$ for any value of n:

$$\sum_{i=1}^{n} i^2 = \frac{n(n+1)(2n+1)}{6}$$

For example, suppose that $n = 7$:

$$\sum_{i=1}^{7} i^2 = 1 + 4 + 9 + 16 + 25 + 36 + 49 = 140$$

The summation formula for square numbers gives you the same answer:

$$\frac{n(n+1)(2n+1)}{6} = \frac{7(8)(15)}{6} = 140$$

The summation formula for cubic numbers

The summation formula for cubic numbers gives you a quick way to add up $1 + 8 + 27 + \ldots + n^3$ for any value of n:

$$\sum_{i=1}^{n} i^3 = \left[\frac{n(n+1)}{2} \right]^2$$

For example, suppose that $n = 5$:

$$\sum_{i=1}^{5} i^3 = 1 + 8 + 27 + 64 + 125 = 225$$

The summation formula for cubic numbers produces the same result:

$$\left[\frac{n(n+1)}{2}\right]^2 = \left[\frac{5(6)}{2}\right]^2 = 15^2 = 225$$

As Bad as It Gets: Calculating Definite Integrals by Using the Riemann Sum Formula

In Chapter 1, I introduce you to this hairy equation for calculating the definite integral:

$$\int_a^b f(x)\,dx = \lim_{n \to \infty} \sum_{i=1}^{n} \left[f\left(x_i^*\right)\left(\frac{b-a}{n}\right) \right]$$

You may be wondering how practical this little gem is for calculating area. That's a valid concern. The bad news is that this formula is, indeed, hairy and you'll need to understand how to use it to pass your first Calculus II exam.

But I have good news, too. In the beginning of Calculus I, you work with an equally hairy equation for calculating derivatives (see Chapter 2 for a refresher). Fortunately, later on, you find a bunch of easier ways to calculate derivatives.

This good news applies to integration, too. Later in this chapter, I show you how to make your life easier. In this section, however, I focus on how to use the Riemann sum formula to calculate the definite integral.

Before I get started, take another look at the Riemann sum formula and notice that the right side of this equation breaks down into four separate "chunks":

- The limit: $\lim_{n \to \infty}$
- The sum: $\sum_{i=1}^{n}$
- The function: $f(x_i^*)$
- The limits of integration: $\frac{b-a}{n}$

To solve an integral by using this formula, work backwards, step by step, as follows:

1. **Plug the limits of integration into the formula.**
2. **Rewrite the function $f(x_i^*)$ as a summation in terms of i and n.**
3. **Calculate the sum.**
4. **Evaluate the limit.**

Plugging in the limits of integration

In this section, I show you how to calculate the following integral:

$$\int_0^4 x^2\, dx$$

This step is a no-brainer: You just plug the limits of integration — that is, the values of a and b — into the formula:

$$\int_0^4 x^2\, dx = \lim_{n \to \infty} \sum_{i=1}^{n}\left[f\left(x_i^*\right)\left(\frac{4-0}{n}\right)\right]$$

Before moving on, I know that you just can't go on living until you simplify $4 - 0$:

$$= \lim_{n \to \infty} \sum_{i=1}^{n}\left[f\left(x_i^*\right)\frac{4}{n}\right]$$

That's it!

Expressing the function as a sum in terms of i and n

This is the tricky step. It's more of an art than a science, so if you're an art major who just happens to be taking a Calculus II course, this just might be your lucky day (or maybe not).

To start out, think about how you would estimate $\int_0^4 x^2\, dx$ by using right rectangles, as I explain earlier in this chapter. Table 3-6 shows you how to do this, using one, two, four, and eight rectangles.

Table 3-6	Using Right Rectangles to Estimate $\int_{0}^{4} x^2\, dx$		
n	**Height**	**Width**	**Expression**
1	4^2	4	$\sum_{i=1}^{1}(4i)^2(4)$
2	$2^2 + 4^2$	2	$\sum_{i=1}^{2}(2i)^2(2)$
4	$1^2 + 2^2 + 3^2 + 4^2$	1	$\sum_{i=1}^{4}i^2(1)$
8	$0.5^2 + 1^2 + 1.5^2 + 2^2 + 2.5^2 + 3^2 + 3.5^2 + 4^2$	0.5	$\sum_{i=1}^{8}(0.5i)^2(0.5)$

Your goal now is to find a *general* expression of the form $\sum_{i=1}^{n}$ that works for every value of n. In the last section, you find that $\frac{4}{n}$ produces the correct width. So, here's the general expression that you're looking for:

$$\sum_{i=1}^{n}\left(\frac{4i}{n}\right)^2\left(\frac{4}{n}\right)$$

Make sure that you understand why this expression works for all values of n before moving on. The first fraction represents the height of the rectangles and the second fraction represents the width, expressed as $\frac{b-a}{n}$.

You can simplify this expression as follows:

$$=\sum_{i=1}^{n}\frac{64i^2}{n^3}$$

Don't forget before moving on that the entire expression is a limit as n approaches infinity:

$$\lim_{n\to\infty}\sum_{i=1}^{n}\frac{64i^2}{n^3}$$

At this point in the problem, you have an expression that's based on two variables: i and n. Remember that the two variables i and n are *in* the sum, and the variable x should already have exited.

Calculating the sum

Now you need a few tricks for calculating the summation portion of this expression:

$$\lim_{n \to \infty} \sum_{i=1}^{n} \frac{64i^2}{n^3}$$

You can ignore the limit in this section — it's just coming along for the ride. You can move a constant outside of a summation without changing the value of that expression:

$$= \lim_{n \to \infty} 64 \sum_{i=1}^{n} \frac{i^2}{n^3}$$

At this point, only the variables i and n are left inside the summation.

Remember that i stands for *icky* and n stands for *nice*. The variable n is nice because you can move it outside the summation just as if it were a constant:

$$= \lim_{n \to \infty} \frac{64}{n^3} \sum_{i=1}^{n} i^2$$

Solving the problem with a summation formula

To handle the icky variable, i, you need a little help. Earlier in the chapter, in "Knowing Sum-Thing about Summation Formulas," I give you some important formulas for handling this summation and others like it.

Getting back to the example, here's where you left off:

$$\lim_{n \to \infty} \frac{64}{n^3} \sum_{i=1}^{n} i^2$$

To evaluate the sum $\sum_{i=1}^{n} i^2$, use the summation formula for square numbers:

$$\lim_{n \to \infty} \frac{64}{n^3} \cdot \frac{n(n+1)(2n+1)}{6}$$

A bit of algebra — which I omit because I know you can do it! — makes the problem look like this:

$$\lim_{n \to \infty} \frac{64}{3} + \frac{1}{n} + \frac{1}{3n^3}$$

You're now set up for the final — and easiest — step.

Evaluating the limit

At this point, the limit that you've probably been dreading all this time turns out to be the simplest part of the problem. As *n* approaches infinity, the two terms with *n* in the denominator approach 0, so they drop out entirely:

$$\lim_{n \to \infty} \frac{64}{3} + \frac{1}{n} + \frac{1}{3n^3} = \frac{64}{3}$$

Yes, this is your final answer! Please note that because you used the Riemann sum formula, this is *not an approximation,* but the *exact* area under the curve $y = x^2$ from 0 to 4.

Light at the End of the Tunnel: The Fundamental Theorem of Calculus

Finding the area under a curve — that is, solving an area problem — can be formalized by using the definite integral (as you discover in Chapter 1). And the definite integral, in turn, is defined in terms of the Riemann sum formula. But, as you find out earlier in this chapter, the Riemann sum formula usually results in lengthy and difficult calculations.

There must be a better way! And, indeed, there is.

The Fundamental Theorem of Calculus (FTC) provides the link between derivatives and integrals. At first glance, these two ideas seem entirely unconnected, so the FTC seems like a bit of mathematical black magic. On closer examination, however, the connection between a function's derivative (its slope) and its integral (the area underneath it) becomes clearer.

In this section, I show you the connection between slope and area. After you see this, the FTC will make more intuitive sense. At that point, I introduce the exact theorem and show you how to use it to evaluate integrals as *antiderivatives* — that is, by understanding integration as the inverse of differentiation.

Without further ado, here's the Fundamental Theorem of Calculus (FTC) in its most useful form:

$$\int_a^b f'(x) = f(b) - f(a)$$

The mainspring of this equality is the connection between *f* and its derivative function *f'*. To solve an integral, you need to be able to *undo* differentiation and find the original function *f*.

Many math books use the following notation for the FTC:

$$\int_a^b f(x)\,dx = F(b) - F(a) \text{ where } F'(x) = f(x)$$

Both notations are equally valid, but I find this version a bit less intuitive than the version that I give you.

The FTC makes evaluating integrals a whole lot easier. For example, suppose that you want to evaluate the following:

$$\int_0^\pi \sin x\,dx$$

This is the function that you see in Figure 3-3. The FTC allows you to solve this problem by thinking about it in a new way. First notice that the following statement is true:

$$f(x) = -\cos x \rightarrow f'(x) = \sin x$$

So the FTC allows you to draw this conclusion:

$$\int_0^\pi \sin x\,dx = (-\cos\pi) - (-\cos 0)$$

Now you can solve this problem by using simple trig:

$$= 1 + 1 = 2$$

So the *exact* (not approximate) shaded area in Figure 3-3 is 2 — all without drawing rectangles! The approximation using the Midpoint Rule (see "Finding a middle ground: The Midpoint Rule" earlier in this chapter) is 2.0944.

As another example, here's the integral that, earlier in the chapter, you solved by using the Riemann sum formula:

$$\int_0^4 x^2\,dx$$

Begin by noticing that the following statement is true:

$$f(x) \frac{1}{3}x^3 \rightarrow f'(x) = x^2$$

Now use the FTC to write this equation:

$$\int_0^4 x^2\,dx = \left(\frac{1}{3}\ 4^3\right) - \left(\frac{1}{3}\ 0^3\right)$$

At this point, the solution becomes a matter of arithmetic:

$$\frac{64}{3} - 0 = \frac{64}{3}$$

In just three simple steps, the definite integral is solved without resorting to the hairy Riemann sum formula!

Understanding the Fundamental Theorem of Calculus

In the previous section, I show you just how useful the Fundamental Theorem of Calculus (FTC) can be for finding the exact value of a definite integral without using the Riemann sum formula. But *why* does the theorem work?

The FTC implies a connection between derivatives and integrals that isn't intuitively obvious. In fact, the theorem implies that derivatives and integrals are inverse operations. It's easy to see why other pairs of operations — such as addition and subtraction — are inverses. But how do you see this same connection between derivatives and integrals?

In this section, I give you a few ways to better understand this connection.

Solving a 200-year-old problem

The connection between derivatives and integrals as inverse operations was first noticed by Isaac Barrow (the teacher of Isaac Newton) in the 17th century. Newton and Gottfried Leibniz (the two key inventors of calculus) both made use of it as a conjecture — that is, as a mathematical statement that's suspected to be true but hasn't been proven yet.

But the FTC wasn't officially proven in all its glory until your old friend Bernhard Riemann demonstrated it in the 19th century. During this 200-year lag, a lot of math — most notably, real analysis — had to be invented before Riemann could prove that derivatives and integrals are inverses.

What's slope got to do with it?

The idea that derivatives and integrals are connected — that is, the slope of a curve and the area under it are linked mathematically — seems odd until you spend some time thinking about it.

If you have a head for business, here's a practical way to understand the connection. Imagine that you own your own company. Envision a graph with a line as your net income (money coming in) and the area under the graph as your net savings (money in the bank). To keep this simple, imagine for the moment that this is a happy world where you have *no* expenses draining your savings account.

When the line on the graph is horizontal, your net income stays the same, so money comes in at a steady rate — that is, your paycheck every week or month is the same. So, your bank account (the area under the line) grows at a steady rate as time passes — that is, as your *x*-value moves to the right.

But suppose that business starts booming. As the line on the graph starts to rise, your paychecks rise proportionally. So, your bank account begins growing at a faster rate.

Now suppose that business slows down. As the line on the graph starts to fall, your paychecks fall proportionally. So, your bank account still grows, but its rate of growth slows down. But beware: If business goes so sour that it can no longer support itself, you may find that you're dipping into savings to support the business, so for the first time your savings goes down.

In this analogy, every *paycheck* is like the area inside a one-unit-wide slice of the graph. And the *bank account* on any particular day is like the total area between the *y*-axis and that day as shown on the graph.

So, when you give it some thought, it would be hard to imagine how slope and area could *not* be connected. The Fundamental Theorem of Calculus is just the exact mathematical representation of this connection.

Introducing the area function

This connection between income (the size of your paycheck) and savings (the amount in your bank account) is a perfect analogy for two important, connected ideas. The income graph represents a function $f(x)$ and the savings graph represents that function's area function $A(x)$.

Figure 3-7 illustrates this connection between $f(x)$ and $A(x)$. This figure represents the *steady income* situation that I describe in the previous section. I choose $f(x) = 1$ to represent income. The resulting savings graph is $A(x) = x$, which rises steadily.

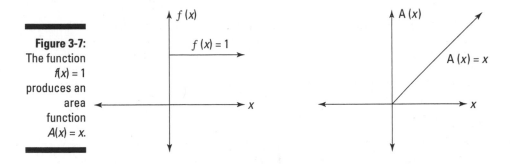

Figure 3-7:
The function
$f(x) = 1$
produces an
area
function
$A(x) = x$.

In comparison, look at Figure 3-8, which represents *rising income*. This time, I choose $f(x) = x$ to represent income. This function produces the area function $A(x) = \frac{1}{2}x^2$, which rises at an increasing rate.

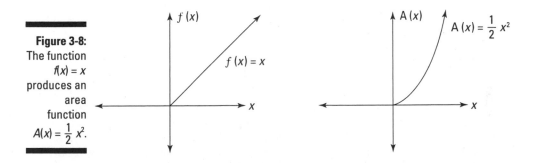

Figure 3-8:
The function
$f(x) = x$
produces an
area
function
$A(x) = \frac{1}{2}x^2$.

Finally, take a peek at Figure 3-9, which represents *falling income*. In this case, I use $f(x) = 2 - x$ to represent income. This function results in the area function $A(x) = 2x - \frac{1}{2}x^2$, which rises at a decreasing rate until the original function drops below 0, and then starts falling.

Take a moment to think about these three examples. Make sure that you see how, in a very practical sense, slope and area are connected: In other words, the slope of a function is the qualitative factor that governs what the related area function looks like.

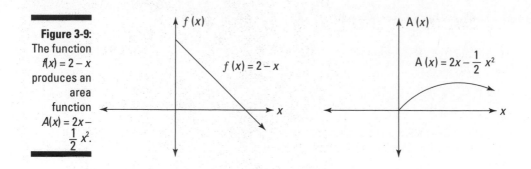

Figure 3-9:
The function
$f(x) = 2 - x$
produces an
area
function
$A(x) = 2x - \frac{1}{2}x^2$.

$f(x)$

$f(x) = 2 - x$

x

$A(x)$

$A(x) = 2x - \frac{1}{2}x^2$

x

Connecting slope and area mathematically

In the previous section, I discuss three functions $f(x)$ and their related area functions $A(x)$. Table 3-7 summarizes this information.

Table 3-7		A Closer Look at Functions and Their Area Functions		
Description of Function	**Equation of Function**	**Description of Area Function**	**Equation of Area Function**	**Derivative of Area Function**
Constant	$f(x) = 1$	Rising steadily	$A(x) = x$	$A'(x) = 1$
Rising	$f(x) = x$	Rising at increasing rate	$A(x) = \frac{1}{2}x^2$	$A'(x) = x$
Falling	$f(x) = 2 - x$	Rising at decreasing rate, and then falling when $f(x) < 0$	$A(x) = 2x - \frac{1}{2}x^2$	$A'(x) = 2 - x$

At this point, the big connection is only a heartbeat away. Notice that each function is the *derivative* of its area function:

$$A'(x) = f(x)$$

Is this mere coincidence? Not at all. Table 3-7 just adds mathematical precision to the intuitive idea that slope of a function (that is, its derivative) is related to the area underneath it.

Because area is mathematically described by the definite integral, as I discuss in Chapter 1, this connection between differentiation and integration makes a whole lot of sense. That's why finding the area under a function — that is, *integration* — is essentially *undoing* a derivative — that is, *anti-differentiation*.

Seeing a dark side of the FTC

Earlier in this chapter, I give you this piece of the Fundamental Theorem of Calculus:

$$\int_a^b f'(x) = f(b) - f(a)$$

Now that you understand the connection between a function $f(x)$ and its area function $A(x)$, here's another piece of the FTC:

$$A_t(x) = \int_s^x f(t)\,dt$$

This piece of the theorem is generally regarded as less useful than the first piece, and it's also harder to grasp because of all the extra variables. I won't belabor it too much, but here are a few points that may help you understand it better:

- The variable s — the lower limit of integration — is an arbitrary starting point where the area function equals zero. In my examples in the previous section, I start the area function at the origin, so $s = 0$. This point represents the day when you opened your bank account, before you deposited any money.

- The variable x — the upper limit of integration — represents any time after you opened your bank account. It's also the independent variable of the area function.

- The variable t is the variable of the function. If you were to draw a graph, t would be the independent variable and $f(t)$ the dependent variable.

In short, don't worry too much about this version of the FTC. The most important thing is that you remember the first version and know how to use it. The other important thing is that you understand how slope and area — that is, derivatives and integrals — are intimately related.

Your New Best Friend: The Indefinite Integral

The Fundamental Theorem of Calculus gives you insight into the connection between a function's slope and the area underneath it — that is, between differentiation and integration.

On a practical level, the FTC gives you an easier way to integrate, without resorting to the Riemann sum formula. This easier way is called *anti-differentiation* — in other words, undoing differentiation. Anti-differentiation is the method that you'll use to integrate throughout the remainder of Calculus II. It leads quickly to a new key concept: the *indefinite integral*.

In this section, I show you step by step how to use the indefinite integral to solve definite integrals, and I introduce the important concept of signed area. To finish the chapter, I make sure that you understand the important distinctions between definite and indefinite integrals.

Introducing anti-differentiation

Integration without resorting to the Riemann sum formula depends upon undoing differentiation (anti-differentiation). Earlier in this chapter, in "Light at the End of the Tunnel: The Fundamental Theorem of Calculus," I calculate a few areas informally by reversing a few differentiation formulas that you know from Calculus I. But anti-differentiation is so important that it deserves its own notation: the indefinite integral.

An *indefinite integral* is simply the notation representing the inverse of the derivative function:

$$\frac{d}{dx} \int f(x)\,dx = f(x)$$

 Be careful not to confuse the indefinite integral with the definite integral. For the moment, notice that the indefinite integral has *no limits of integration*. Later in this chapter, in "Distinguishing definite and indefinite integrals," I outline the differences between these two types of integrals.

Here are a few examples that informally connect derivatives that you know with indefinite integrals that you want to be able to solve:

$$\frac{d}{dx} \sin x = \cos x \rightarrow \int \cos x\,dx = \sin x$$

$$\frac{d}{dx} e^x = e^x \rightarrow \int e^x\,dx = e^x$$

$$\frac{d}{dx} \ln |x| = \frac{1}{x} \rightarrow \int \frac{1}{x}\,dx = \ln x$$

There's a small but important catch in this informal analysis. Notice that the following three statements are all true:

$$\frac{d}{dx} \sin x + 1 = \cos x$$

$$\frac{d}{dx} \sin x - 100 = \cos x$$

$$\frac{d}{dx} \sin x + 1{,}000{,}000 = \cos x$$

Because any constant differentiates to 0, you need to account for the possible presence of a constant when integrating. So, here are the more precise formulations of the indefinite integrals I just introduced:

$$\int \cos x \, dx = \sin x + C$$

$$\int e^x \, dx = e^x + C$$

$$\int \frac{1}{x} \, dx = \ln x + C$$

The formal solution of every indefinite integral is an anti-derivative up to the addition of a constant C, which is called the *constant of integration*. So, just mechanically attach a + C whenever you evaluate an indefinite integral.

Solving area problems without the Riemann sum formula

After you know how to solve an indefinite integral by using anti-differentiation (as I show you in the previous section), you have at your disposal a very useful method for solving area problems. This announcement should come as a great relief, especially after reading the earlier section "As Bad as It Gets: Calculating Definite Integrals by Using the Riemann Sum Formula."

Here's how you solve an area problem by using indefinite integrals — that is, *without* resorting to the Riemann sum formula:

1. **Formulate the area problem as a definite integral (as I show you in Chapter 1).**

2. **Solve the definite integral as an indefinite integral evaluated between the given limits of integration.**

3. **Plug the limits of integration into this expression and simplify to find the area.**

This method is, in fact, the one that you use for solving area problems for the rest of Calculus II. For example, suppose that you want to find the shaded area in Figure 3-10.

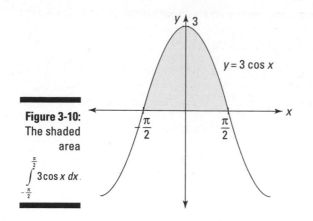

Here's how you do it:

1. **Formulate the area problem as a definite integral:**

$$\int_{-\frac{\pi}{2}}^{\frac{\pi}{2}} 3\cos x\, dx$$

2. **Solve this definite integral as an indefinite integral:**

$$= 3\sin x\Big|_{x=-\frac{\pi}{2}}^{x=\frac{\pi}{2}}$$

I replace the integral with the expression $3\sin x$, because $\frac{d}{dx}\, 3\sin x = 3\cos x$. I also introduce the notation $\Big|_{x=-\frac{\pi}{2}}^{x=\frac{\pi}{2}}$. You can read it as *evaluated from x equals* $-\frac{\pi}{2}$ *to x equals* $\frac{\pi}{2}$. This notation is commonly used so that you can show your teacher that you know how to integrate and postpone worrying about the limits of integration until the next step.

3. **Plug these limits of integration into the expression and simplify:**

$$= 3\sin\frac{\pi}{2} - 3\sin -\frac{\pi}{2}$$

As you can see, this step comes straight from the FTC, subtracting $f(b) - f(a)$. Now, I just simplify this expression to find the area:

$$= 3 - (-3) = 6$$

So, the area of the shaded region in Figure 3-10 equals 6.

No *C*, no problem!

You may wonder why the constant of integration *C* — which is so important when you're evaluating an indefinite integral — gets dropped when you're evaluating a definite integral. This one is easy to explain.

Remember that every definite integral is expressed as the difference between a function evaluated at one point and *the same function* evaluated at another point. If this function includes a constant *C*, one *C* cancels out the other.

For example, take the definite integral $\int_{0}^{\frac{\pi}{6}} \cos x \, dx$. Technically speaking, this integral is evaluated as follows:

$$\sin x + c \Big|_{x=0}^{x=\frac{\pi}{6}}$$

$$= (\sin \frac{\pi}{6} + C) - (\sin 0 + C)$$

$$= \frac{1}{2} + C - 0 - C = \frac{1}{2}$$

As you can clearly "*C*," the two *C*s cancel each other out, so there's no harm in dropping them at the beginning of the evaluation rather than at the end.

Understanding signed area

In the real world, the smallest possible area is 0, so area is always a nonnegative number. On the graph, however, area can be either positive or negative.

This idea of negative area relates back to a discussion earlier in this chapter, in "Introducing the area function," where I talk about what happens when a function dips below the *x*-axis.

To use the analogy of income and savings, this is the moment when your income dries up and money starts flowing out. In other words, you're spending your savings, so your savings account balance starts to fall.

So, area above the *x*-axis is positive, but area below the *x*-axis is measured as negative area.

The definite integral takes this important distinction into account. It provides not just the area but the *signed area* of a region on the graph. For example, suppose that you want to measure the shaded area in Figure 3-11.

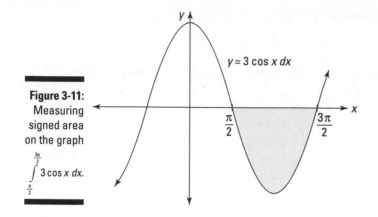

Here's how you do it using the steps that I outline in the previous section:

1. **Formulate the area problem as a definite integral:**

$$\int_{\frac{\pi}{2}}^{\frac{3\pi}{2}} 3\cos x \, dx$$

2. **Solve this definite integral as an indefinite integral:**

$$= 3\sin x \Big|_{x=\frac{\pi}{2}}^{x=\frac{3\pi}{2}}$$

3. **Plug these limits of integration into the expression and simplify:**

$$= 3\sin\frac{3\pi}{2} - 3\sin\frac{\pi}{2}$$
$$= -3 - 3 = -6$$

So, the signed area of the shaded region in Figure 3-11 equals –6. As you can see, the computational method for evaluating the definite integral gives the signed area automatically.

As another example, suppose that you want to find the total area of the two shaded regions in Figure 3-10 and Figure 3-11. Here's how you do it using the steps that I outline in the previous section:

1. **Formulate the area problem as a definite integral:**

$$\int_{-\frac{\pi}{2}}^{\frac{3\pi}{2}} \cos x \, dx$$

2. **Solve this definite integral as an indefinite integral:**

$$= 3\sin x \Big|_{x=-\frac{\pi}{2}}^{x=\frac{3\pi}{2}}$$

3. **Plug these limits of integration into the expression and simplify:**

$$= -3 \sin \frac{3\pi}{2} - 3 \sin \frac{\pi}{2}$$

$$= 3 - 3 = 0$$

This time, the signed area of the shaded region is 0. This answer makes sense, because the unsigned area above the x-axis equals the unsigned area below it, so these two areas cancel each other out.

Distinguishing definite and indefinite integrals

Don't confuse the definite and indefinite integrals. Here are the key differences between them:

A definite integral

- ✔ Includes limits of integration (a and b)
- ✔ Represents the exact area of a specific set of points on a graph
- ✔ Evaluates to a number

An indefinite integral

- ✔ Doesn't include limits of integration
- ✔ Can be used to evaluate an infinite number of related definite integrals
- ✔ Evaluates to a function

For example, here's a *definite* integral:

$$\int_{0}^{\frac{\pi}{4}} \sec^2 x \, dx$$

As you can see, it includes limits of integration (0 and $\frac{\pi}{4}$), so you can draw a graph of the area that it represents. You can then use a variety of methods to evaluate this integral as a *number*. This number equals the signed area between the function and the x-axis inside the limits of integration, as I discuss earlier in "Understanding signed area."

In contrast, here's an *indefinite* integral:

$$\int \sec^2 x \, dx$$

This time, the integral doesn't include limits of integration, so it doesn't represent a specific area. Thus, it doesn't evaluate to a number, but to a function:

$$= \tan x + C$$

You can use this function to evaluate any related definite integral. For example, here's how to use it to evaluate the definite integral I just gave you:

$$\int_0^{\frac{\pi}{4}} \sec^2 x \, dx$$

$$= \tan x \Big|_{x=0}^{x=\frac{\pi}{4}}$$

$$= 2 \tan \frac{\pi}{4} - \tan 0$$

$$= 1 - 0 = 1$$

So, the area of the shaded region in the graph is 1.

As you can see, the indefinite integral encapsulates an infinite number of related definite integrals. It also provides a practical means for evaluating definite integrals. Small wonder that much of Calculus II focuses on evaluating indefinite integrals. In Part II, I give you an ordered approach to evaluating indefinite integrals.

Part II
Indefinite Integrals

The 5th Wave   By Rich Tennant

"It's 'Fast Herschel Fenniman', the most notorious math
hustler of all time. If he asks if you'd like to run some
trigonometric integrals with him, just walk away."

In this part . . .

You begin calculating the indefinite integral as an anti-derivative — that is, as the inverse of a derivative. In practice, this is easier for some functions than others. So, I show you four important tricks — variable substitution, integration by parts, trig substitution, and integrating with partial fractions — for turning a function you don't know how to integrate into one that you do.

Chapter 4

Instant Integration: Just Add Water (And C)

In This Chapter

▶ Calculating simple integrals as anti-derivatives

▶ Using 17 integral formulas and 3 integration rules

▶ Integrating more difficult functions by using more than one integration tool

▶ Clarifying the difference between integrative and nonintegrable functions

F irst the good news: Because integration is the inverse of differentiation, you already know how to evaluate a lot of basic integrals.

Now the bad news: In practice, integration is often a lot trickier than differentiation. I'm telling you this upfront because a) it's true; b) I believe in honesty; and c) you should prepare yourself before your first exam. (Buying and reading this book, by the way, are *great* first steps!)

In this chapter — and also in Chapters 5 through 8 — I focus exclusively on one question: How do you integrate every single function on the planet? Okay, I'm exaggerating, but not by much. I give you a manageable set of integration techniques that you can do with a pencil and paper, and if you know when and how to apply them, you'll be able to integrate everything but the kitchen sink.

First, I show you how to start integrating by thinking about integration as anti-differentiation — that is, as the inverse of differentiation. I give you a not-too-long list of basic integrals, which mirrors the list of basic derivatives from Chapter 2. I also give you a few rules for breaking down functions into manageable chunks that are easier to integrate.

After that, I show you a few techniques for tweaking functions to make them look like the functions you already know how to integrate. By the end of this chapter, you have the tools to integrate dozens of functions quickly and easily.

Evaluating Basic Integrals

In Calculus I (which I cover in Chapter 2), you find that a few algorithms — such as the Product Rule, Quotient Rule, and Chain Rule — give you the tools to differentiate just about every function your professor could possibly throw at you. In Calculus II, students often greet the news that "there's no Chain Rule for integration" with celebratory cheers. By the middle of the semester, they usually revise this opinion.

Using the 17 basic anti-derivatives for integrating

In Chapter 2, I give you a list of 17 derivatives to know, cherish, and above all *memorize* (yes, I said *memorize*). Reading that list may lead you to believe that I'm one of those harsh über-math dudes who takes pleasure in cruel and unusual curricular activities.

But math is kind of like the Ghost of Christmas Past — the stuff you thought was long ago dead and buried comes back to haunt you. And so it is with derivatives. If you already know them, you'll find this section easy.

The Fundamental Theorem of Calculus shows that integration is the inverse of differentiation up to a constant C. This key theorem gives you a way to begin integrating. In Table 4-1, I show you how to integrate a variety of common functions by identifying them as the derivatives of functions you already know.

Table 4-1	The 17 Basic Integrals (Anti-Derivatives)
Derivative	*Integral (Anti-Derivative)*
$\frac{d}{dx} \, n = 0$	$\int 0 \, dx = C$
$\frac{d}{dx} \, x = 1$	$\int 1 \, dx = x + C$
$\frac{d}{dx} \, e^x = e^x$	$\int e^x = e^x + C$
$\frac{d}{dx} \, \ln x = \frac{1}{x}$	$\int \frac{1}{x} \, dx = \ln x + C$
$\frac{d}{dx} \, n^x = n^x \ln n$	$\int n^x \, dx = \frac{n^x}{\ln n} + C$
$\frac{d}{dx} \, \sin x = \cos x$	$\int \cos x \, dx = \sin x + C$

Derivative	Integral (Anti-Derivative)
$\dfrac{d}{dx}\cos x = -\sin x$	$\displaystyle\int \sin x\, dx = -\cos x + C$
$\dfrac{d}{dx}\tan x = \sec^2 x$	$\displaystyle\int \sec^2 x\, dx = \tan x + C$
$\dfrac{d}{dx}\cot x = -\csc^2 x$	$\displaystyle\int \csc^2 x\, dx = -\cot x + C$
$\dfrac{d}{dx}\sec x = \sec x \tan x$	$\displaystyle\int \sec x \tan x\, dx = \sec x + C$
$\dfrac{d}{dx}\csc x = -\csc x \cot x$	$\displaystyle\int \csc x \cot x\, dx = -\csc x + C$
$\dfrac{d}{dx}\arcsin x = \dfrac{1}{\sqrt{1-x^2}}$	$\displaystyle\int \dfrac{1}{\sqrt{1-x^2}}\, dx = \arcsin x + C$
$\dfrac{d}{dx}\arccos x = \dfrac{-1}{\sqrt{1-x^2}}$	$\displaystyle\int -\dfrac{1}{\sqrt{1-x^2}}\, dx = \arccos x + C$
$\dfrac{d}{dx}\arctan x = \dfrac{1}{1+x^2}$	$\displaystyle\int \dfrac{1}{1+x^2}\, dx = \arctan x + C$
$\dfrac{d}{dx}\operatorname{arccot} x = -\dfrac{1}{1+x^2}$	$\displaystyle\int -\dfrac{1}{1+x}\, dx = \operatorname{arccot} x + C$
$\dfrac{d}{dx}\operatorname{arcsec} x = \dfrac{1}{x\sqrt{x^2-1}}$	$\displaystyle\int \dfrac{1}{x\sqrt{x^2-1}}\, dx = \operatorname{arcsec} x + C$
$\dfrac{d}{dx}\operatorname{arccsc} x = -\dfrac{1}{x\sqrt{x^2-1}}$	$\displaystyle\int -\dfrac{1}{x\sqrt{x^2-1}} = \operatorname{arccsc} x + C$

As I discuss in Chapter 3, you need to add the constant of integration C because constants differentiate to 0. For example:

$$\frac{d}{dx}\sin x = \cos x$$

$$\frac{d}{dx}\sin x + 1 = \cos x$$

$$\frac{d}{dx}\sin x - 100 = \cos x$$

So when you integrate by using anti-differentiation, you need to account for the potential presence of this constant:

$$\int \cos x\, dx = \sin x + C$$

Three important integration rules

After you know how to integrate by using the 17 basic anti-derivatives in Table 4-1, you can expand your repertoire with three additional integration rules: the Sum Rule, the Constant Multiple Rule, and the Power Rule. These three rules mirror those that you know from differentiation.

The Sum Rule for integration

The Sum Rule for integration tells you that integrating long expressions term by term is okay. Here it is formally:

$$\int \left[f(x) + g(x) \right] dx = \int f(x)\,dx + \int g(x)\,dx$$

For example:

$$\int \left(\cos x + x^2 - \frac{1}{x} \right) dx = \int \cos x\,dx + \int x^2\,dx - \int \frac{1}{x}\,dx$$

Note that the Sum Rule also applies to expressions of more than two terms. It also applies regardless of whether the term is positive or negative. (Some books call this variation the Difference Rule, but you get the idea.) Splitting this integral into three parts allows you to integrate each separately by using a different anti-differentiation rule:

$$= \sin x + \frac{1}{3} x^3 - \ln x + C$$

Notice that I add only one C at the end. Technically speaking, you should add one variable of integration (say, C_1, C_2, and C_3) for each integral that you evaluate. But, at the end, you can still declare the variable $C = C_1 + C_2 + C_3$ to consolidate all these variables. In most cases when you use the Sum Rule, you can skip this step and just tack a C onto the end of the answer.

The Constant Multiple Rule for integration

The Constant Multiple Rule tells you that you can move a constant outside of a derivative before you integrate. Here it is expressed in symbols:

$$\int nf(x)\,dx = n \int f(x)\,dx$$

For example:

$$\int 3 \tan x \sec x\,dx = 3 \int \tan x \sec x\,dx$$

As you can see, this rule mirrors the Constant Multiple Rule for differentiation. With the constant out of the way, integrating is now easy using an anti-differentiation rule:

$$= 3 \sec x + C$$

The Power Rule for integration

The Power Rule for integration allows you to integrate any real power of x (except –1). Here's the Power Rule expressed formally:

$$\int x^n \, dx = \frac{1}{n+1} x^{n+1} + C$$

For example:

$$\int x \, dx = \frac{1}{2} x^2 + C$$

$$\int x^2 \, dx = \frac{1}{3} x^3 + C$$

$$\int x^{100} \, dx = \frac{1}{101} x^{101} + C$$

The Power Rule works fine for negative powers of x, which are powers of x in the denominator. For example:

$$\int \frac{1}{x^2} \, dx$$

$$= \int x^{-2} \, dx$$

$$= -x^{-1} + C$$

$$= -\frac{1}{x} + C$$

The Power Rule also works for rational powers of x, which are roots of x. For example:

$$\int \sqrt{x^3} \, dx$$

$$= \int x^{\frac{3}{2}} \, dx$$

$$= \frac{2}{5} x^{\frac{5}{2}} + C$$

$$= \frac{2}{5} \sqrt{x^5} + C$$

The *only* real-number power that the Power Rule doesn't work for is –1. Fortunately, you have an anti-differentiation rule to handle this case:

$$\int \frac{1}{x} \, dx$$

$$= \int x^{-1} \, dx$$

$$= \ln |x| + C$$

What happened to the other rules?

Integration contains formulas that mirror the Sum Rule, the Constant Multiple Rule, and the Power Rule for differentiation. But it lacks formulas that look like the Product Rule, Quotient Rule, and Chain Rule. This fact may sound like good news, but the lack of formulas makes integration a lot trickier in practice than differentiation is.

In fact, Chapters 5 through 8 focus on a bunch of methods that mathematicians have devised for getting around this difficulty. Chapter 5 focuses on variable substitution, which is a limited form of the Chain Rule. And in Chapter 6, I show you integration by parts, which is an adaptation of the Product Rule.

Evaluating More Difficult Integrals

The anti-differentiation rules for integrating, which I explain earlier in this chapter, greatly limit how many integrals you can compute easily. In many cases, however, you can tweak a function to make it easier to integrate.

In this section, I show you how to integrate certain fractions and roots by using the Power Rule. I also show you how to use the trig identities in Chapter 2 to stretch your capacity to integrate trig functions.

Integrating polynomials

You can integrate *any* polynomial in three steps by using the rules from this section:

1. **Use the Sum Rule to break the polynomial into its terms and integrate each of these separately.**

2. **Use the Constant Multiple Rule to move the coefficient of each term outside its respective integral.**

3. **Use the Power Rule to evaluate each integral. (You only need to add a single *C* to the end of the resulting expression.)**

For example, suppose that you want to evaluate the following integral:

$$\int \left(10x^6 - 3x^3 + 2x - 5\right) dx$$

1. **Break the expression into four separate integrals:**

$$= \int 10x^6 \, dx - \int 3x^3 \, dx + \int 2x \, dx - \int 5 \, dx$$

2. **Move each of the four coefficients outside its respective integral:**

$$= 10 \int x^6 \, dx - 3 \int x^3 \, dx + 2 \int x \, dx - 5 \int dx$$

3. **Integrate each term separately using the Power Rule:**

$$= \frac{10}{7}x^7 - \frac{3}{4}x^4 + x^2 - 5x + C$$

You can integrate *any* polynomial by using this method. Many integration methods I introduce later in this book rely on this fact. So, practice integrating polynomials until you feel so comfortable that you could do it in your sleep.

Integrating rational expressions

In many cases, you can untangle hairy rational expressions and integrate them by using the anti-differentiation rules plus the other three rules in this chapter.

For example, here's an integral that looks like it may be difficult:

$$\int \frac{(x^2+5)(x-3)^2}{\sqrt{x}} \, dx$$

You can split the function into several fractions, but without the Product Rule or Quotient Rule, you're then stuck. Instead, expand the numerator and put the denominator in exponential form:

$$= \int \frac{x^4 - 6x^3 + 14x^2 - 30x + 45}{x^{\frac{1}{2}}} \, dx$$

Next, split the expression into five terms:

$$= \int \left(x^{\frac{7}{2}} - 6x^{\frac{5}{2}} + 14x^{\frac{3}{2}} - 30x^{\frac{1}{2}} + 45x^{-\frac{1}{2}} \right) dx$$

Then, use the Sum Rule to separate the integral into five separate integrals and the Constant Multiple Rule to move the coefficient outside the integral in each case:

$$= \int x^{\frac{7}{2}} \, dx - 6 \int x^{\frac{5}{2}} \, dx + 14 \int x^{\frac{3}{2}} \, dx - 30 \int x^{\frac{1}{2}} \, dx + 45 \int x^{-\frac{1}{2}} \, dx$$

Now, you can integrate each term separately using the Power Rule:

$$= \frac{2}{9}x^{\frac{9}{2}} - \frac{12}{7}x^{\frac{7}{2}} + \frac{28}{5}x^{\frac{5}{2}} - 20x^{\frac{3}{2}} + 90x^{\frac{1}{2}} + C$$

Using identities to integrate trig functions

At first glance, some products or quotients of trig functions may seem impossible to integrate by using the formulas I give you earlier in this chapter. But, you'll be surprised how much headway you can often make when you integrate an unfamiliar trig function by first tweaking it using the Basic Five trig identities that I list in Chapter 2.

The unseen power of these identities lies in the fact that they allow you to express *any* combination of trig functions into a combination of sines and cosines. Generally speaking, the trick is to simplify an unfamiliar trig function and turn it into something that you know how to integrate.

When you're faced with an unfamiliar product or quotient of trig functions, follow these steps:

1. **Use trig identities to turn all factors into sines and cosines.**

2. **Cancel factors wherever possible.**

3. **If necessary, use trig identities to eliminate all fractions.**

For example:

$$\int \sin^2 x \cot x \sec x \, dx$$

In its current form, you can't integrate this expression by using the rules from this chapter. So you follow these steps to turn it into an expression you can integrate:

1. **Use the identities** $\cot x = \frac{\cos x}{\sin x}$ **and** $\sec x = \frac{1}{\cos x}$:

$$= \int \sin^2 x \cdot \frac{\cos x}{\sin x} \cdot \frac{1}{\cos x} \, dx$$

2. **Cancel both sin x and cos x in the numerator and denominator:**

$$= \int \sin x \, dx$$

In this example, even without Step 3, you have a function that you can integrate.

$$= -\cos x + C$$

Here's another example:

$$\int \tan x \sec x \csc x \, dx$$

Again, this integral looks like a dead end before you apply the five basic trig identities to it:

1. **Turn all three factors into sines and cosines:**

   $$= \int \frac{\sin x}{\cos x} \cdot \frac{1}{\cos x} \cdot \frac{1}{\sin x}\, dx$$

2. **Cancel sin x in the numerator and denominator:**

   $$= \int \frac{1}{\cos^2 x}\, dx$$

3. **Use the identity $\cos x = \frac{1}{\sec x}$ to eliminate the fraction:**

   $$= \int \sec^2 x\, dx$$

   $$= \tan x + C$$

Again, you turn an unfamiliar function into one of the ten trig functions that you know how to integrate.

I show you lots more tricks for integrating trig functions in Chapter 7.

Understanding Integrability

By now, you've probably figured out that, in practice, integration is usually harder than differentiation. The lack of any set rules for integrating products, quotients, and compositions of functions makes integration something of an art rather than a science.

So, you may think that a large number of functions are differentiable, with a smaller subset of these being integrable. It turns out that this conclusion is false. In fact, the set of integrable functions is larger, with a smaller subset of these being differentiable. To understand this fact, you need to be clear on what the words *integrable* and *differentiable* really mean.

In this section, I shine some light on two common mistakes that students make when trying to understand what integrability is all about. After that, I discuss what it means for a function to be integrable, and I show you why many functions that are integrable aren't differentiable.

Understanding two red herrings of integrability

In trying to understand what makes a function integrable, you first need to understand two related issues: difficulties in *computing integrals* and *representing integrals as functions*. These issues are valid, but they're red herrings — that is, they don't really affect whether a function is integrable.

Computing integrals

For many input functions, integrals are more difficult to compute than derivatives are. For example, suppose that you want to differentiate and integrate the following function:

$$y = 3x^5 e^{2x}$$

You can differentiate this function easily by using the Product Rule (I take an additional step to simplify the answer):

$$\frac{dy}{dx} = 3\left[\frac{d}{dx}\left(x^5\right)e^{2x} + \frac{d}{dx}\left(e^{2x}\right)x^5\right]$$
$$= 3(5x^4 e^{2x} + 2e^{2x}x^5)$$
$$= 3x^4 e^{2x}(2x + 5)$$

Because no such rule exists for integration, in this example you're forced to seek another method. (You find this method in Chapter 6, where I discuss integration by parts.)

Finding solutions to integrals can be tricky business. In comparison, finding derivatives is comparatively simple — you learned most of what you need to know about it in Calculus I.

Representing integrals as functions

Beyond difficulties in computation, the integrals of certain functions simply can't be represented by using the functions that you're used to.

More precisely, some integrals can't be represented as *elementary functions* — that is, as combinations of the functions you know from Pre-Calculus. (See Chapter 14 for a more in-depth look at elementary functions.)

For example, take the following function:

$$y = e^{x^2}$$

You can find the derivative of the function easily by using the Chain Rule:

$$\frac{d}{dx}e^{x^2}$$
$$=e^{x^2}\left(\frac{d}{dx}x^2\right)$$
$$=e^{x^2}(2x)$$
$$=2xe^{x^2}$$

However, the integral of the same function, e^{x^2}, can't be expressed as a function — at least, not any function that you're used to.

Instead, you can express this integral either *exactly* — as an infinite series — or *approximately* — as a function that approximates the integral to a given level of precision. (See Part IV for more on infinite series.) Alternatively, you can just leave it as an integral, which also expresses it just fine for some purposes.

Understanding what integrable really means

When mathematicians discuss whether a function is integrable, they aren't talking about the difficulty of computing that integral — or even whether a method has been discovered. Each year, mathematicians find new ways to integrate classes of functions. However, this fact doesn't mean that previously nonintegrable functions are now integrable.

Similarly, a function's integrability also doesn't hinge upon whether its integral can be easily represented as another function, without resorting to infinite series.

In fact, when mathematicians say that a function is integrable, they mean only that the integral is *well defined* — that is, that it makes mathematical sense.

In practical terms, integrability hinges on continuity: If a function is continuous on a given interval, it's integrable on that interval. Additionally, if a function has only a finite number of discontinuities on an interval, it's also integrable on that interval.

You probably remember from Calculus I that many functions — such as those with discontinuities, sharp turns, and vertical slopes — are nondifferentiable. Discontinuous functions are also nonintegrable. However, functions with sharp turns and vertical slopes are integrable.

For example, the function $y = |x|$ contains a sharp point at $x = 0$, so the function is nondifferentiable at this point. However, the same function is integrable for all values of x. This is just one of infinitely many examples of a function that's integrable but not differentiable in the entire set of real numbers.

So, surprisingly, the set of differentiable functions is actually a subset of the set of integrable functions. In practice, however, computing the integral of most functions is more difficult than computing the derivative.

Chapter 5

Making a Fast Switch: Variable Substitution

*U*nlike differentiation, integration doesn't have a Chain Rule. This fact makes integrating *compositions of functions* (functions within functions) a little bit tricky. The most useful trick for integrating certain common compositions of functions uses variable substitution.

With *variable substitution,* you set a variable (usually *u*) equal to part of the function that you're trying to integrate. The result is a simplified function that you can integrate by using the anti-differentiation formulas and the three basic integration rules (Sum Rule, Constant Multiple Rule, and Power Rule — all discussed in Chapter 4).

In this chapter, I show you how to use variable substitution. Then I show you how to identify a few common situations where variable substitution is helpful. After you get comfortable with the process, I give you a quick way to integrate by just looking at the problem and writing down the answer. Finally, I show you how to skip a step when using variable substitution to evaluate definite integrals.

Knowing How to Use Variable Substitution

The anti-differentiation formulas plus the Sum Rule, Constant Multiple Rule, and Power Rule (all discussed in Chapter 4) allow you to integrate a variety of common functions. But as functions begin to get a little bit more complex, these methods become insufficient. For example, these methods don't work on the following:

$$\int \sin 2x \, dx$$

To evaluate this integral, you need some stronger medicine. The sticking point here is the presence of the constant 2 inside the sine function. You have an anti-differentiation rule for integrating the sine of a variable, but how do you integrate the sine of a variable times a constant?

The answer is variable substitution, a five-step process that allows you to integrate where no integral has gone before:

1. **Declare a variable *u* and set it equal to an algebraic expression that appears in the integral, and then substitute *u* for this expression in the integral.**

2. **Differentiate *u* to find $\frac{du}{dx}$, and then isolate all *x* variables on one side of the equal sign.**

3. **Make another substitution to change *dx* and all other occurrences of *x* in the integral to an expression that includes *du*.**

4. **Integrate by using *u* as your new variable of integration.**

5. **Express this answer in terms of *x*.**

I don't expect these steps to make much sense until you see how they work in action. In the rest of this section, I show you how to use variable substitution to solve problems that you wouldn't be able to integrate otherwise.

Finding the integral of nested functions

Suppose that you want to integrate the following:

$$\int \sin 2x \, dx$$

The difficulty here lies in the fact that this function is the composition of two functions: the function $2x$ nested inside a sine function. If you were differentiating, you could use the Chain Rule. Unfortunately, no Chain Rule exists for integration.

Fortunately, this function is a good candidate for variable substitution. Follow the five steps I give you in the previous section:

1. **Declare a new variable u as follows and substitute it into the integral:**

   Let $u = 2x$

   Now, substitute u for $2x$ as follows:

   $$\int \sin 2x \, dx = \int \sin u \, dx$$

   This may look like the answer to all your troubles, but you have one more problem to resolve. As it stands, the symbol dx tells you that variable of integration is still x.

   To integrate properly, you need to find a way to change dx to an expression containing du. That's what Steps 2 and 3 are about.

2. **Differentiate the function $u = 2x$ and isolate the x terms on one side of the equal sign:**

   $$\frac{du}{dx} = 2$$

   Now, treat the symbol $\frac{du}{dx}$ as if it's a fraction, and isolate the x terms on one side of the equal sign. I do this in two steps:

   $$du = 2 \, dx$$
   $$\frac{1}{2} \, du = dx$$

3. **Substitute $\frac{1}{2} du$ for dx into the integral:**

   $$\int \sin u \left(\frac{1}{2} \, du \right)$$

   You can treat the $\frac{1}{2}$ just like any coefficient and use the Constant Multiple Rule to bring it outside the integral:

   $$= \frac{1}{2} \int \sin u \, du$$

4. **At this point, you have an expression that you know how to evaluate:**

   $$= -\frac{1}{2} \cos u + C$$

5. Now that the integration is done, the last step is to substitute $2x$ back in for u:

$$-\frac{1}{2} \cos 2x + C$$

You can check this solution by differentiating using the Chain Rule:

$$\frac{d}{dx} \left(-\frac{1}{2} \cos 2x + C\right)$$

$$= \frac{d}{dx}\left(-\frac{1}{2} \cos 2x\right) + \frac{d}{dx} C$$

$$= -\frac{1}{2} (-\sin 2x) (2) + 0$$

$$= \sin 2x$$

Finding the integral of a product

Imagine that you're faced with this integral:

$$\int \sin^3 x \cos x \, dx$$

The problem in this case is that the function that you're trying to integrate is the product of two functions — $\sin^3 x$ and $\cos x$. This would be simple to differentiate with the Product Rule, but integration doesn't have a Product Rule. Again, variable substitution comes to the rescue:

1. Declare a variable as follows and substitute it into the integral:

Let $u = \sin x$

You may ask *how* I know to declare u equal to $\sin x$ (rather than, say, $\sin^3 x$ or $\cos x$). I answer this question later in the chapter. For now, just follow along and get the mechanics of variable substitution.

You can substitute this variable into the expression that you want to integrate as follows:

$$\int \sin^3 x \cos x \, dx = \int u^3 \cos x \, dx$$

Notice that the expression $\cos x \, dx$ still remains and needs to be expressed in terms of u.

2. Differentiate the function $u = \sin x$ and isolate the x variables on one side of the equal sign:

$$\frac{du}{dx} = \cos x$$

Isolate the x variables on one side of the equal sign:

$$du = \cos x \, dx$$

3. **Substitute du for $\cos x\, dx$ in the integral:**

$$\int u^3\, du$$

4. **Now you have an expression that you can integrate:**

$$= \frac{1}{4} u^4 + C$$

5. **Substitute $\sin x$ for u:**

$$= \frac{1}{4} \sin^4 x + C$$

And again, you can check this answer by differentiating with the Chain Rule:

$$\frac{d}{dx} \left(\frac{1}{4} \sin^4 x + C\right)$$

$$= \frac{d}{dx} \frac{1}{4} \sin^4 x + \frac{d}{dx} C$$

$$= \frac{1}{4} (4 \sin^3 x)(\cos x) + 0$$

$$= \sin^3 x \cos x$$

This derivative matches the original function, so the integration is correct.

Integrating a function multiplied by a set of nested functions

Suppose that you want to integrate the following:

$$\int x \sqrt{3x^2 + 7}\, dx$$

This time, you're trying to integrate the product of a function (x) and a composition of functions (the function $3x^2 + 7$ nested inside a square root function). If you were differentiating, you could use a combination of the Product Rule and the Chain Rule, but these options aren't available for integration. Here's how you integrate, step by step, by using variable substitution:

1. **Declare a variable u as follows and substitute it into the integral:**

   Let $u = 3x^2 + 7$

   Here, you may ask how I know what value to assign to u. Here's the short answer: u is the inner function, as you would identify if you were using the Chain Rule. (See Chapter 2 for a review of the Chain Rule.) I explain this more fully later in "Recognizing When to Use Substitution."

   Now, substitute u into the integral:

   $$\int x \sqrt{3x^2 + 7}\, dx = \int x \sqrt{u}\, dx$$

Make one more small rearrangement to place all the remaining x terms together:

$$= \int \sqrt{u}\, x \, dx$$

This rearrangement makes clear that I still have to find a substitution for $x\, dx$.

2. **Now differentiate the function $u = 3x^2 + 7$:**

$$\frac{du}{dx} = 6x$$

From Step 1, I know that I need to replace $x\, dx$ in the integral:

$$du = 6x\, dx$$

$$\frac{1}{6}\, du = x\, dx$$

3. **Substitute $\dfrac{du}{6}$ for $x\, dx$:**

$$= \int \sqrt{u}\left(\frac{1}{6}\, du\right)$$

You can move the fraction $\frac{1}{6}$ outside the integral:

$$= \frac{1}{6} \int \sqrt{u}\, du$$

4. **Now you have an integral that you know how to evaluate.**

I take an extra step, putting the square root in exponential form, to make sure that you see how to do this:

$$= \frac{1}{6} \int u^{\frac{1}{2}}\, du$$

$$= \frac{1}{6}\left(\frac{2}{3}\right) u^{\frac{3}{2}} + C$$

$$= \frac{1}{9}\, u^{\frac{3}{2}} + C$$

5. **To finish up, substitute $3x^2 + 7$ for u:**

$$= \frac{1}{9}\left(3x^2 + 7\right)^{\frac{3}{2}} + C$$

As with the first two examples in this chapter, you can always check your integration by differentiating the result:

$$\frac{d}{dx}\left[\frac{1}{9}\left(3x^2 + 7\right)^{\frac{3}{2}} + C\right]$$

$$= \frac{d}{dx}\frac{1}{9}\left(3x^2 + 7\right)^{\frac{3}{2}} + \frac{d}{dx}C$$

$$= \frac{1}{9}\left(\frac{3}{2}\right)\left(3x^2 + 7\right)^{\frac{1}{2}}(6x) + 0$$

$$= x\sqrt{3x^2 + 7}$$

As if by magic, the derivative brings you back to the function you started with.

Recognizing When to Use Substitution

In the previous section, I show you the mechanics of variable substitution — that is, *how* to perform variable substitution. In this section, I clarify *when* to use variable substitution.

You may be able to use variable substitution in three common situations. In these situations, the expression you want to evaluate is one of the following:

- A composition of functions — that is, a function nested in a function
- A function multiplied by a function
- A function multiplied by a computation of functions

Integrating nested functions

Compositions of functions — that is, one function nested inside another — are of the form $f(g(x))$. You can integrate them by substituting $u = g(x)$ when

- You know how to integrate the outer function f.
- The inner function $g(x)$ differentiates to a constant — that is, it's of the form ax or $ax + b$.

Example #1

Here's an example. Suppose that you want to integrate the function

$$\csc^2 (4x + 1) \, dx$$

Again, this is a composition of two functions:

- The outer function f is the $\csc^2$ function, which you know how to integrate.
- The inner function is $g(x) = 4x + 1$, which differentiates to the constant 4.

This time the composition is held together by the equality $u = 4x + 1$. That is, the two basic functions $f(u) = \csc^2 u$ and $g(x) = 4x + 1$ are composed by the equality $u = 4x + 1$ to produce the function $f(g(x)) = \csc^2 (4x + 1)$.

Both criteria are met, so this integral is another prime candidate for substitution using $u = 4x + 1$. Here's how you do it:

1. **Declare a variable u and substitute it into the integral:**

    Let $u = 4x + 1$

    $$\int \csc^2(4x + 1) \, dx = \int \csc^2 u \, dx$$

2. Differentiate $u = 4x + 1$ and isolate the x term:

$$\frac{du}{dx} = 4$$

$$\frac{du}{4} = dx$$

3. Substitute $\frac{du}{4}$ for dx in the integral:

$$\int \csc^2 u \left(\frac{1}{4} \, du \right)$$

$$= \frac{1}{4} \int \csc^2 u \, du$$

4. Evaluate the integral:

$$= -\frac{1}{4} \cot u + C$$

5. Substitute back $4x + 1$ for u:

$$= -\frac{1}{4} \cot (4x + 1) + C$$

Example #2

Here's one more example. Suppose that you want to evaluate the following integral:

$$\int \frac{1}{x - 3} \, dx$$

Again, this is a composition of two functions:

- ✔ The outer function f is a fraction — technically, an exponent of –1 — which you know how to integrate.
- ✔ The inner function is $g(x) = x - 3$, which differentiates to 1.

Here, the composition is held together by the equality $u = x - 3$. That is, the two basic functions $f(u) = \frac{1}{u}$ and $g(x) = x - 3$ are composed by the equality $u = x - 3$ to produce the function $f(g(x)) = \frac{1}{x - 3}$.

The criteria are met, so you can integrate by using the equality $u = x - 3$:

1. Declare a variable u and substitute it into the integral:

Let $u = x - 3$

$$\int \frac{1}{x - 3} \, dx = \int \frac{1}{u} \, dx$$

2. Differentiate $u = x - 3$ and isolate the x term:

$$\frac{du}{dx} = 1$$

$$du = dx$$

3. **Substitute** *du* **for** *dx* **in the integral:**

   $\int \frac{1}{u} du$

4. **Evaluate the integral:**

   $= \ln |u| + C$

5. **Substitute back** *x* − 3 **for** *u***:**

   $= \ln |x - 3| + C$

Knowing a shortcut for nested functions

After you work through enough examples of variable substitution, you may begin to notice certain patterns emerging. As you get more comfortable with the concept, you can use a shortcut to integrate compositions of functions — that is, nested functions of the form $f(g(x))$. Technically, you're using the variable substitution $u = g(x)$, but you can bypass this step and still get the right answer.

This shortcut works for compositions of functions $f(g(x))$ for which

- ✔ You know how to integrate the outer function *f*.
- ✔ The inner function $g(x)$ is of the form *ax* or *ax* + *b* — that is, it differentiates to a constant.

When these two conditions hold, you can integrate $f(g(x))$ by using the following three steps:

1. **Write down the reciprocal of the coefficient of** *x***.**

2. **Multiply by the integral of the outer function, copying the inner function as you would when using the Chain Rule in differentiation.**

3. **Add** *C***.**

Example #1

For example:

$\int \cos 4x \, dx$

Notice that this is a function nested within a function, where the following are true:

- ✔ The outer function *f* is the cosine function, which you know how to integrate.
- ✔ The inner function is $g(x) = 4x$, which is of the form *ax*.

So, you can integrate this function quickly as follows:

1. **Write down the reciprocal of 4 — that is, $\frac{1}{4}$:**

   $$\frac{1}{4}$$

2. **Multiply this reciprocal by the integral of the outer function, copying the inner function:**

   $$\frac{1}{4} \sin 4x$$

3. **Add C:**

   $$\frac{1}{4} \sin 4x + C$$

That's it! You can check this easily by differentiating, using the Chain Rule:

$$\frac{d}{dx}\left(\frac{1}{4} \sin 4x + C\right)$$

$$= \frac{1}{4} \cos 4x \,(4)$$

$$= \cos 4x$$

Example #2

Here's another example:

$$\int \sec^2 10x\, dx$$

Remember as you begin that $\sec^2 10x\, dx$ is a notational shorthand for $[\sec(10x)]^2$. So, the outer function f is the $\sec^2$ function and the inner function is $g(x) = 10x$. (See Chapter 2 for more on the ins and outs of trig notation.) Again, the criteria for variable substitution are met:

1. **Write down the reciprocal of 10 — that is, $\frac{1}{10}$:**

   $$\frac{1}{10}$$

2. **Multiply this reciprocal by the integral of the outer function, copying the inner function:**

   $$\frac{1}{10} \tan 10x$$

3. **Add C:**

   $$\frac{1}{10} \tan 10x + C$$

Here's the check:

$$\frac{d}{dx}\left(\frac{1}{10}\tan 10x + C\right)$$

$$=\frac{d}{dx}\frac{1}{10}\tan 10x + \frac{d}{dx}C$$

$$=\frac{1}{10}\sec^2 10x\,(10) + 0$$

$$=\sec^2 10x$$

Example #3

Here's another example:

$$\int \frac{1}{7x + 2}\,dx$$

In this case, the outer function is division, which counts as a function, as I explain earlier in "Recognizing When to Use Substitution." The inner function is $7x + 2$. Both of these functions meet the criteria, so here's how to perform this integration:

1. **Write down the reciprocal of the coefficient 7 — that is, $\frac{1}{7}$:**

   $$\frac{1}{7}$$

2. **Multiply this reciprocal by the integral of the outer function, copying the inner function:**

   $$\frac{1}{7}\ln|7x + 2|$$

3. **Add C:**

   $$\frac{1}{7}\ln|7x + 2| + C$$

You're done! As always, you can check your result by differentiating, using the Chain Rule:

$$\frac{d}{dx}\left(\frac{1}{7}\ln|7x + 2| + C\right)$$

$$=\frac{1}{7}\left(\frac{1}{7x + 2}\right)(7)$$

$$=\frac{1}{7x + 2}$$

Example #4

Here's one more example:

$$\int \sqrt{12x - 5}\, dx$$

This time, the outer function f is a square root — that is, an exponent of $\frac{1}{2}$ — and $g(x) = 12x - 5$, so you can use a quick substitution:

1. **Write down the reciprocal of 12 — that is, $\frac{1}{12}$:**

$$\frac{1}{12}$$

2. **Multiply the integral of the outer function, copying down the inner function:**

$$\frac{1}{12}\frac{2}{3}(12x - 5)^{\frac{3}{2}}$$

$$= \frac{1}{18}(12x - 5)^{\frac{3}{2}}$$

3. **Add C:**

$$\frac{1}{18}(12x - 5)^{\frac{3}{2}} + C$$

Table 5-1 gives you a variety of integrals in this form. As you look over this chart, get a sense of the pattern so that you can spot it when you have an opportunity to integrate quickly.

Table 5-1	Using the Shortcut for Integrating Nested Functions
Integral	**Evaluation**
$\int e^{5x}\, dx$	$\frac{1}{5} e^{5x} + C$
$\int \sin 7x\, dx$	$-\frac{1}{7} \cos 7x + C$
$\int \sec^2 \frac{x}{3}\, dx$	$3 \tan \frac{x}{3} + C$
$\int \tan 8x \sec 8x\, dx$	$\frac{1}{8} \sec 8x + C$
$\int e^{5x+2}\, dx$	$\frac{1}{5} e^{5x+2} + C$
$\int \cos(x - 4)\, dx$	$\sin(x - 4) + C$

Substitution when one part of a function differentiates to the other part

When $g'(x) = f(x)$, you can use the substitution $u = g(x)$ to integrate the following:

✔ Expressions of the form $f(x) \cdot g(x)$

✔ Expressions of the form $f(x) \cdot h(g(x))$, provided that h is a function that you already know how to integrate

Don't worry if you don't understand all this math-ese. In the following sections, I show you how to recognize both of these cases and integrate each. As usual, variable substitution helps to fill the gaps left by the absence of a Product Rule and a Chain Rule for integration.

Expressions of the form $f(x) \cdot g(x)$

Some products of functions yield quite well to variable substitution. Look for expressions of the form $f(x) \cdot g(x)$ where

✔ You know how to integrate $g(x)$.

✔ The function $f(x)$ is the derivative of $g(x)$.

For example:

$$\int \tan x \sec^2 x \, dx$$

The main thing to notice here is that the derivative of $\tan x$ is $\sec^2 x$. This is a great opportunity to use variable substitution:

1. **Declare u and substitute it into the integral:**

   Let $u = \tan x$

   $$\int \tan x \sec^2 x \, dx = \int u \sec^2 x \, dx$$

2. **Differentiate as planned:**

   $$\frac{du}{dx} = \sec^2 x$$
   $$du = \sec^2 x \, dx$$

3. **Perform another substitution:**

   $$= \int u \, du$$

4. **This integration couldn't be much easier:**

$$= \frac{1}{2} u^2 + C$$

5. **Substitute back tan x for u:**

$$= \frac{1}{2} \tan^2 x + C$$

Expressions of the form $f(x) \cdot h(g(x))$

Here's a hairy-looking integral that actually responds well to substitution:

$$\int \frac{(2x+1)}{\left(x^2 tx - 5\right)^{\frac{4}{3}}} \, dx$$

The key insight here is that the numerator of this fraction is the derivative of the inner function in the denominator. Watch how this plays out in this substitution:

1. **Declare u equal to the denominator and make the substitution:**

    Let $u = x^2 + x - 5$

    Here's the substitution:

    $$= \int \frac{2x+1}{u^{\frac{4}{3}}} \, dx$$

2. **Differentiate u:**

    $$\frac{du}{dx} = 2x + 1$$

    $$du = (2x + 1) \, dx$$

3. **The second part of the substitution now becomes clear:**

    $$= \int \frac{1}{u^{\frac{4}{3}}} \, du$$

    Notice how this substitution hinges on the fact that the numerator is the derivative of the denominator. (You may think that this is quite a coincidence, but coincidences like these happen all the time on exams!)

4. **Integration is now quite straightforward:**

    I take an extra step to remove the fraction before I integrate.

    $$= \int u^{-\frac{4}{3}} \, du$$

    $$= -3u^{-\frac{1}{3}} + C$$

5. **Substitute back $x^2 + x - 5$ for u:**

    $$= -3(x^2 + x - 5)^{-\frac{1}{3}} + C$$

Checking the answer by differentiating with the Chain Rule reveals how this problem was set up in the first place:

$$\frac{d}{dx}\left[-3(x^2 + x - 5)^{-\frac{1}{3}} + C\right]$$
$$= (x^2 + x - 5)^{-\frac{4}{3}}(2x + 1)$$
$$= \frac{2x + 1}{\left(x^2 + x - 5\right)^{\frac{4}{3}}}$$

By now, if you've worked through the examples in this chapter, you're probably seeing opportunities to make variable substitutions. For example:

$$\int x^3 \sqrt{x^4 - 1}\, dx$$

Notice that the derivative of $x^4 - 1$ is x^3, off by a constant factor. So here's the declaration, followed by the differentiation:

Let $u = x^4 - 1$
$$\frac{du}{dx} = 4x^3$$
$$\frac{du}{4} = x^3\, dx$$

Now you can just do both substitutions at once:

$$\int \sqrt{u} \cdot \left(\frac{1}{4}\, du\right)$$
$$= \frac{1}{4} \int \sqrt{u}\, du$$

At this point, you can solve the integral simply — I'll leave this as an exercise for you!

Similarly, here's another example:

$$\int \csc^2 x\, e^{\cot x}\, dx$$

At first glance, this integral looks just plain horrible. But on further inspection, notice that the derivative of $\cot x$ is $-\csc^2 x$, so this looks like another good candidate:

Let $u = \cot x$
$$\frac{du}{dx} = -\csc^2 x$$
$$-du = \csc^2 x\, dx$$

This results in the following substitution:

$$= \int e^u (-du)$$

$$= -\int e^u \, du$$

Again, this is another integral that you can solve.

Using Substitution to Evaluate Definite Integrals

In the first two sections of this chapter, I cover how and when to evaluate indefinite integrals with variable substitution. All this information also applies to evaluating definite integrals, but I also have a timesaving trick that you should know.

When using variable substitution to evaluate a definite integral, you can save yourself some trouble at the end of the problem. Specifically, you can leave the solution in terms of u by changing the limits of integration.

For example, suppose that you're evaluating the following definite integral:

$$\int_{x=0}^{x=1} x\sqrt{x^2 + 1} \, dx$$

Notice that I give the limits of integration as $x = 0$ and $x = 1$. This is just a notational change to remind you that the limits of integration are values of x. This fact becomes important later in the problem.

You can evaluate this equation simply by using variable substitution.

If you're not sure why this substitution works, read the section "Recognizing When to Use Substitution" earlier in this chapter. Follow Steps 1 through 3 of variable substitution:

Let $u = x^2 + 1$

$$\frac{du}{dx} = 2x$$

$$\frac{du}{2} = x \, dx$$

$$= \frac{1}{2} \int_{x=0}^{x=2} \sqrt{u} \, du$$

If this were an indefinite integral, you'd be ready to integrate. But because this is a definite integral, you still need to express the limits of integration in terms of u rather than x. Do this by substituting values 0 and 1 for x in the substitution equation $u = x^2 + 1$:

$$u = 1^2 + 1 = 2$$
$$u = 0^2 + 1 = 1$$

Now use these values of u as your new limits of integration:

$$= \frac{1}{2} \int_{v=1}^{v=2} \sqrt{u}\, du$$

At this point, you're ready to integrate:

$$= \frac{1}{2} \cdot \frac{2}{3}\, u^{\frac{3}{2}} \Big|_{u=1}^{u=2}$$

$$= \frac{1}{3}\, u^{\frac{3}{2}} \Big|_{v=1}^{v=2}$$

Because you changed the limits of integration, you can now find the answer without switching the variable back to x:

$$= \frac{1}{3} \left(2^{\frac{3}{2}} - 1^{\frac{3}{2}} \right)$$

$$= \frac{1}{3} \left(\sqrt{8} - 1 \right)$$

$$= \frac{\sqrt{8}}{3} - \frac{1}{3}$$

Chapter 6

Integration by Parts

- -

In This Chapter

▶ Making the connection between the Product Rule and integration by parts

▶ Knowing how and when integration by parts works

▶ Integrating by parts by using the DI-agonal method

▶ Practicing the DI-agonal method on the four most common products of functions

- -

*I*n Calculus I, you find that the Product Rule allows you to find the derivative of any two functions that are multiplied together. (I review this in Chapter 2, in case you need a refresher.) But integrating the product of two functions isn't quite as simple. Unfortunately, no formula allows you to integrate the product of any two functions. As a result, a variety of techniques have been developed to handle products of functions on a case-by-case basis.

In this chapter, I show you the most widely applicable technique for integrating products, called *integration by parts*. First, I demonstrate how the formula for integration by parts follows the Product Rule. Then I show you how the formula works in practice. After that, I give you a list of the products of functions that are likely to yield to this method.

After you understand the principle behind integration by parts, I give you a method — called the *DI-agonal method* — for performing this calculation efficiently and without errors. Then I show you examples of how to use this method to integrate the four most common products of functions.

Introducing Integration by Parts

Integration by parts is a happy consequence of the Product Rule (discussed in Chapter 2). In this section, I show you how to tweak the Product Rule to derive the formula for integration by parts. I show you two versions of this formula — a complicated version and a simpler one — and then recommend that you memorize the second. I show you how to use this formula, and then I give you a heads up as to when integration by parts is likely to work best.

Reversing the Product Rule

The Product Rule (see Chapter 2) enables you to differentiate the product of two functions:

$$\frac{d}{dx}\left[f(x) \cdot g(x)\right] = f'(x) \cdot g(x) + g'(x) \cdot f(x)$$

Through a series of mathematical somersaults, you can turn this equation into a formula that's useful for integrating. This derivation doesn't have any truly difficult steps, but the notation along the way is mind-deadening, so don't worry if you have trouble following it. Knowing how to derive the formula for integration by parts is less important than knowing when and how to use it, which I focus on in the rest of this chapter.

The first step is simple: Just rearrange the two products on the right side of the equation:

$$\frac{d}{dx}\left[f(x) \cdot g(x)\right] = f(x) \cdot g'(x) + g(x) \cdot f'(x)$$

Next, rearrange the terms of the equation:

$$f(x) \cdot g'(x) = \frac{d}{dx}\left[f(x) \cdot g(x)\right] - g(x) \cdot f'(x)$$

Now, integrate both sides of this equation:

$$\int f(x)g'(x)\,dx = \int \left\{ \frac{d}{dx}\left[f(x)g(x)\right] - g(x)f'(x) \right\} dx$$

Use the Sum Rule to split the integral on the left in two:

$$\int f(x)g'(x)\,dx = \int \frac{d}{dx}\left[f(x)g(x)\right] dx - \int g(x)f'(x)\,dx$$

The first of these two integrals undoes the derivative:

$$\int f(x)g'(x)\,dx = f(x)g(x) - \int g(x)f'(x)\,dx$$

This is the formula for integration by parts. But because it's so hairy looking, the following substitution is used to simplify it:

Let $u = f(x)$ Let $v = g(x)$

$du = f'(x)\,dx$ $dv = g'(x)\,dx$

Here's the friendlier version of the same formula, which you should *memorize:*

$$\int u\,dv = uv - \int v\,du$$

Knowing how to integrate by parts

The formula for integration by parts gives you the option to break the product of two functions down to its factors and integrate it in an altered form.

To integrate by parts:

1. **Decompose the entire integral (including *dx*) into two factors.**
2. **Let the factor without *dx* equal *u* and the factor with *dx* equal *dv*.**
3. **Differentiate *u* to find *du*, and integrate *dv* to find *v*.**
4. **Use the formula $\int u\,dv = uv - \int v\,du$.**
5. **Evaluate the right side of this equation to solve the integral.**

For example, suppose that you want to evaluate this integral:

$$\int x\ln x\,dx$$

In its current form, you can't perform this computation, so integrate by parts:

1. **Decompose the integral into ln *x* and *x dx*.**
2. **Let *u* = ln *x* and *dv* = *x dx*.**
3. **Differentiate ln *x* to find *du* and integrate *x dx* to find *v*:**

   Let $u = \ln x$ Let $dv = x\,dx$

   $\dfrac{du}{dx} = \dfrac{1}{x}$ $\int dv = \int x\,dx$

   $du = \dfrac{1}{x}\,dx$ $v = \dfrac{1}{2}x^2$

4. **Using these values for *u*, *du*, *v*, and *dv*, you can use the formula for integration by parts to rewrite the integral as follows:**

   $$\int x\ln x\,dx = (\ln x)\left(\frac{1}{2}x^2\right) - \int \left(\frac{1}{2}x^2\right)\left(\frac{1}{x}\right)dx$$

   At this point, algebra is useful to simplify the right side of the equation:

   $$= \frac{1}{2}x^2\ln x - \frac{1}{2}\int x\,dx$$

5. **Evaluate the integral on the right:**

   $$= \frac{1}{2}x^2\ln x - \frac{1}{2}\left(\frac{1}{2}\right)x^2 + C$$

   You can simplify this answer just a bit:

   $$= \frac{1}{2}x^2\ln x - \frac{1}{4}x^2 + C$$

Therefore, $\int x \ln x \, dx$. To check this answer, differentiate it by using the Product Rule:

$$\frac{d}{dx} \left(\frac{1}{2} x^2 \ln x - \frac{1}{4} x^2 + C \right)$$

$$= \frac{1}{2} \left[\left(\frac{d}{dx} x^2 \right) \ln x + \left(\frac{d}{dx} \ln x \right) x^2 \right] - \frac{1}{4} 2x$$

$$= \frac{1}{2} \left[2x \ln x + \left(\frac{1}{x} \cdot x^2 \right) \right] - \frac{1}{2} x$$

Now, simplify this result to show that it's equivalent to the function you started with:

$$= x \ln x + \frac{1}{2} x - \frac{1}{2} x = x \ln x$$

Knowing when to integrate by parts

After you know the basic mechanics of integrating by parts, as I show you in the previous section, it's important to recognize when integrating by parts is useful.

To start off, here are two important cases when integration by parts is definitely the way to go:

- ✔ The logarithmic function $\ln x$
- ✔ The first four inverse trig functions (arcsin x, arccos x, arctan x, and arccot x)

Beyond these cases, integration by parts is useful for integrating the product of more than one function. For example:

- ✔ $x \ln x$
- ✔ $x \operatorname{arcsec} x$
- ✔ $x^2 \sin x$
- ✔ $e^x \cos x$

Notice that in each case, you can recognize the product of functions because the variable x appears more than once in the function.

Whenever you're faced with integrating the product of functions, consider variable substitution (which I discuss in Chapter 5) before you think about integration by parts. For example, $x \cos (x^2)$ is a job for variable substitution, not integration by parts. (To see why, flip to Chapter 5.)

When you decide to use integration by parts, your next question is how to split up the function and assign the variables u and dv. Fortunately, a helpful mnemonic exists to make this decision: <u>L</u>ovely <u>I</u>ntegrals <u>A</u>re <u>T</u>errific, which stands for <u>L</u>ogarithmic, <u>I</u>nverse trig, <u>A</u>lgebraic, <u>T</u>rig. (If you prefer, you can also use the mnemonic <u>L</u>ousy <u>I</u>ntegrals <u>A</u>re <u>T</u>errible.) Always choose the *first* function in this list as the factor to set equal to u, and then set the rest of the product (including dx) equal to dv.

You can use integration by parts to integrate any of the functions listed in Table 6-1.

Table 6-1	When You Can Integrate by Parts		
Function	*Example*	*Differentiate u to Find du*	*Integrate dv to Find v*
Log function	$\int \ln x \, dx$	$\ln x$	dx
Log times algebraic	$\int x^4 \ln x \, dx$	$\ln x$	$x^4 \, dx$
Log composed with algebraic	$\int \ln x^3 \, dx$	$\ln x^3$	dx
Inverse trig forms	$\int \arcsin x \, dx$	$\arcsin x$	dx
Algebraic times sine	$\int x^2 \sin x \, dx$	x^2	$\sin x \, dx$
Algebraic times cosine	$\int 3x^5 \cos x \, dx$	$3x^5$	$\sin x \, dx$
Algebraic times exponential	$\int \frac{1}{2} x^2 e^{3x} \, dx$	$\frac{1}{2} x^2$	$e^3 \, dx$
Sine times exponential	$\int e^{\frac{x}{2}} \sin x \, dx$	$e^{\frac{x}{2}}$	$\sin x \, dx$
Cosine times exponential	$\int e^x \cos x \, dx$	e^x	$\cos x \, dx$

When you're integrating by parts, here's the most basic rule when deciding which term to integrate and which to differentiate: If you only know how to integrate one of the two, that's the one you integrate!

Integrating by Parts with the DI-agonal Method

The DI-agonal method is basically integration by parts with a chart that helps you organize information. This method is especially useful when you need to integrate by parts more than once to solve a problem. In this section, I show you how to use the DI-agonal method to evaluate a variety of integrals.

Looking at the DI-agonal chart

The DI-agonal method avoids using u and dv, which are easily confused (especially if you write the letters u and v as sloppily as I do!). Instead, a column for *differentiation* is used in place of u, and a column for *integration* replaces dv.

Use the following chart for the DI-agonal method:

	I
D	
+	
−	

As you can see, the chart contains two columns: the D column for *differentiation,* which has a plus sign and a minus sign, and the I column for *integration.* You may also notice that the D and the I are placed *diagonally* in the chart — yes, the name *DI-agonal method* works on two levels (so to speak).

Using the DI-agonal method

Earlier in this chapter, I provide a list of functions that you can integrate by parts. The DI-agonal method works for all these functions. I also give you the mnemonic **L**ovely **I**ntegrals **A**re **T**errific (which stands for **L**ogarithmic, **I**nverse trig, **A**lgebraic, **T**rig) to help you remember how to assign values of u and dv — that is, what to differentiate and what to integrate.

To use the DI-agonal method:

1. **Write the value to differentiate in the box below the D and the value to integrate (omitting the dx) in the box below the I.**

2. **Differentiate down the D column and integrate down the I column.**

3. **Add the products of all *full* rows as terms.**

   I explain this step in further detail in the examples that follow.

4. **Add the integral of the product of the two lowest diagonally adjacent boxes.**

   I also explain this step in greater detail in the examples.

Don't spend too much time trying to figure this out. The upcoming examples show you how it's done and give you plenty of practice. I show you how to use the DI-agonal method to integrate products that include logarithmic, inverse trig, algebraic, and trig functions.

L is for logarithm

You can use the DI-agonal method to evaluate the product of a log function and an algebraic function. For example, suppose that you want to evaluate the following integral:

$$\int x^2 \ln x \, dx$$

Whenever you integrate a product that includes a log function, the log function always goes in the D column.

1. **Write the log function in the box below the D and the rest of the function value (omitting the dx) in the box below the I.**

	I
D	x^2
$+$ $\ln x$	
$-$	

2. **Differentiate ln x and place the answer in the D column.**

   Notice that in this step, the minus sign already in the box attaches to $\frac{1}{x}$.

	I
D	x^2
+ ln x	
- $\frac{1}{x}$	

3. Integrate x^2 and place the answer in the *I* column.

	I
D	x^2
+ ln x	$\frac{1}{3}x^3$
- $\frac{1}{x}$	

4. Add the product of the full row that's circled.

	I
D	x^2
+ ln x	$\frac{1}{3}x^3$
- $\frac{1}{x}$	

Here's what you write:

$$+\ln x \left(\frac{1}{3}x^3 \right)$$

5. Add the *integral* of the two lowest diagonally adjacent boxes that are circled.

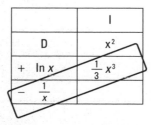

Here's what you write:

$$(+\ln x)\left(\frac{1}{3}x^3\right) + \int\left(-\frac{1}{x}\right)\left(\frac{1}{3}x^3\right)dx$$

At this point, you can simplify the first term and integrate the second term:

$$=\frac{1}{3}x^3\ln x - \frac{1}{3}\int x^2\,dx$$

$$=\frac{1}{3}x^3\ln x - \left(\frac{1}{3}\right)\left(\frac{1}{3}x^3\right) + C$$

$$=\frac{1}{3}x^3\ln x - \frac{1}{9}x^3 + C$$

You can verify this answer by differentiating by using the Product Rule:

$$\frac{d}{dx}\left(\frac{1}{3}x^3\ln x - \frac{1}{9}x^3 + C\right)$$

$$=\frac{1}{3}\left(3x^2\ln x + \frac{1}{x}\cdot x^3\right) - \frac{1}{3}x^2$$

$$=x^2\ln x + \frac{1}{3}x^2 - \frac{1}{3}x^2$$

$$=x^2\ln x$$

Therefore, this is the correct answer:

$$\int x^2\ln x\,dx = \frac{1}{3}x^3\ln x - \frac{1}{9}x^3 + C$$

I is for inverse trig

You can integrate four of the six inverse trig functions (arcsin x, arccos x, arctan x, and arccot x) by using the DI-agonal method. By the way, if you haven't memorized the derivatives of the six inverse trig functions (which I give you in Chapter 2), this would be a great time to do so.

Whenever you integrate a product that includes an inverse trig function, this function always goes in the D column.

For example, suppose that you want to integrate

$$\int \arccos x\,dx$$

1. **Write the inverse trig function in the box below the D and the rest of the function value (omitting the dx) in the box below the I.**

	I
D	1
+ arccos x	
−	

   Note that a 1 goes into the I column.

2. **Differentiate arccos x and place the answer in the D column, and then integrate 1 and place the answer in the I column.**

	I
D	1
+ arccos x	x
$-\left(-\dfrac{1}{\sqrt{1-x^2}}\right)$	

3. **Add the product of the full row that's circled.**

	I
D	1
+ arccos x	x
$-\left(-\dfrac{1}{\sqrt{1-x^2}}\right)$	

   Here's what you write:

   $(+\text{arccos } x)(x)$

4. Add the integral of the lowest diagonal that's circled.

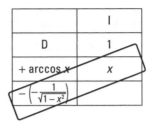

Here's what you write:

$$(+\arccos x)(x) + \int -\left(-\frac{1}{\sqrt{1-x^2}}\right)(x)\,dx$$

Simplify and integrate:

$$=x\arccos x + \int \frac{x}{\sqrt{1-x^2}}\,dx$$

Let $u = 1 - x^2$

$du = -2x\,dx$

$-\dfrac{1}{2}\,du = x\,dx$

This variable substitution introduces a new variable u. Don't confuse this u with the u used for integration by parts.

$$=x\arccos x + \int \frac{1}{\sqrt{u}}\left(-\frac{1}{2}\,du\right)$$

$$=x\arccos x - \frac{1}{2}\left(2\sqrt{u}\right) + C$$

$$=x\,\arccos x - \sqrt{u} + C$$

Substituting $1 - x^2$ for u and simplifying gives you this answer:

$$=x\arccos x - \sqrt{1-x^2} + C$$

Therefore, $\int \arccos x\,dx = x\arccos x - \sqrt{1-x^2} + C$.

A is for algebraic

If you're a bit skeptical that the DI-agonal method is really worth the trouble, I guarantee you that you'll find it useful when handling algebraic factors.

For example, suppose that you want to integrate the following:

$$\int x^3 \sin x \, dx$$

This example is a product of functions, so integration by parts is an option. Going down the LIAT checklist, you notice that the product doesn't contain a log factor or an inverse trig factor. But it does include the algebraic factor x^3, so place this factor in the D column and the rest in the I column. By now, you're probably getting good at using the chart, so I fill it in for you here.

		I
	D	sin x
+	x^3	$-$ cos x
$-$	$3x^2$	

Your next step is normally to write the following

$$+(x^3)(-\cos x) + \int (-3x^2)(-\cos x) \, dx$$

But here comes trouble: The only way to calculate the new integral is by doing *another* integration by parts. And, peeking ahead a bit, here's what you have to look forward to:

$$= (x^3)(-\cos x) - \left[(3x^2)(-\sin x) - \int (6x)(-\sin x) \, dx\right]$$

$$= (x^3)(-\cos x) - \left\{(3x^2)(-\sin x) - \left[(6x)(\cos x) - \int 6\cos x \, dx\right]\right\}$$

At last, after integrating by parts *three times,* you finally have an integral that you can solve directly. If evaluating this expression looks like fun (and if you think you can do it quickly on an exam without dropping a minus sign along the way!), by all means go for it. If not, I show you a better way. Read on.

To integrate an algebraic function multiplied by a sine, a cosine, or an exponential function, place the algebraic factor in the D column and the other factor in the I column. Differentiate the algebraic factor down to zero, and then integrate the other factor the same number of times. You can then copy the answer directly from the chart.

Simply extend the DI chart as I show you here.

	I
D	sin x
+ x^3	– cos x
– $3x^2$	– sin x
+ $6x$	cos x
– 6	sin x
+ 0	

Notice that you just continue the patterns in both columns. In the D column, continue alternating plus and minus signs and differentiate until you reach 0. And in the I column, continue integrating.

The very pleasant surprise is that you can now copy the answer from the chart. This answer contains four terms (+ C, of course), which I copy directly from the four circled rows in the chart:

$$x^3 (-\cos x) - 3x^2 (-\sin x) + 6x (\cos x) - 6 (\sin x) + C$$

But wait! Didn't I forget the final integral on the diagonal? Actually, no — but this integral is $\int 0\, dx \cdot \sin x = C$, which explains where that final C comes from.

Here's another example, just to show you again how easy the DI-agonal method is for products with algebraic factors:

$$\int 3x^5 e^{2x}\, dx$$

Without the DI chart, this problem is one gigantic miscalculation waiting to happen. But the chart keeps track of everything.

	I
D	e^{2x}
+ $3x^5$	$\frac{1}{2}\, e^{2x}$
− $15x^4$	$\frac{1}{4}\, e^{2x}$
+ $60x^3$	$\frac{1}{8}\, e^{2x}$
− $180x^2$	$\frac{1}{16}\, e^{2x}$
+ $360x$	$\frac{1}{32}\, e^{2x}$
− 360	$\frac{1}{64}\, e^{2x}$
+ 0	

Now, just copy from the chart, add C, and simplify:

$$= +(3x^5)\left(\frac{1}{2}\,e^{2x}\right) - (15x^4)\left(\frac{1}{4}\,e^{2x}\right) + (60x^3)\left(\frac{1}{8}\,e^{2x}\right) - (180x^2)\left(\frac{1}{16}\,e^{2x}\right)$$
$$+ (360x)\left(\frac{1}{32}\,e^{2x}\right) - (360)\left(\frac{1}{64}\,e^{2x}\right) + C$$
$$= \frac{3}{2}\,x^5 e^{2x} - \frac{15}{4}\,x^4 e^{2x} + \frac{15}{2}\,x^3 e^{2x} - \frac{45}{4}\,x^2 e^{2x} + \frac{45}{4}\,x e^{2x} - \frac{45}{8}\,e^{2x} + C$$

This answer is perfectly acceptable, but if you want to get fancy, factor out ⅜ e^{2x} and leave a reduced polynomial:

$$= \frac{3}{8}\,e^{2x}\,(4x^5 - 10x^4 + 20x^3 - 30x^2 + 30x - 15) + C$$

T is for trig

You can use the DI-agonal method to integrate the product of either a sine or a cosine and an exponential. For example, suppose that you want to evaluate the following integral:

$$\int e^{\frac{x}{3}} \sin x \, dx$$

When integrating either a sine or cosine function multiplied by an exponential function, make your DI-agonal chart with five rows rather than four. Then place the trig function in the D column and the exponential in the I column.

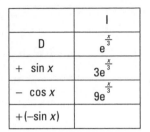

D	I
	$e^{\frac{x}{3}}$
+ $\sin x$	$3e^{\frac{x}{3}}$
− $\cos x$	$9e^{\frac{x}{3}}$
+(−$\sin x$)	

This time, you have two rows to add, as well as the integral of the product of the lowest diagonal:

$$(\sin x)\left(3e^{\frac{x}{3}}\right) + (-\cos x)\left(9e^{\frac{x}{3}}\right) + \int (-\sin x)\left(9e^{\frac{x}{3}}\right)dx$$

This may seem like a dead end because the resulting integral looks so similar to the one that you're trying to evaluate. Oddly enough, however, this similarity makes solving the integral possible. In fact, the next step is to make the integral that results look *exactly* like the one you're trying to solve:

$$=(\sin x)\left(3e^{\frac{x}{3}}\right) + (-\cos x)\left(9e^{\frac{x}{3}}\right) - 9\int e^{\frac{x}{3}}\sin x\, dx$$

Next, substitute the variable I for the integral that you're trying to solve. This action isn't strictly necessary, but it makes the course of action a little clearer.

$$I = (\sin x)(3e^{\frac{x}{3}}) + (-\cos x)(9e^{\frac{x}{3}}) - 9I$$

Now solve for I using a little basic algebra:

$$10I = (\sin x)(3e^{\frac{x}{3}}) + (-\cos x)(9e^{\frac{x}{3}})$$

$$I = \frac{(\sin x)\left(3e^{\frac{x}{3}}\right) + (-\cos x)\left(9e^{\frac{x}{3}}\right)}{10}$$

Finally, substitute the original integral back into the equation, and add C:

$$\int e^{\frac{x}{3}}\sin x\, dx = \frac{1}{10}\left[(\sin x)\left(3e^{\frac{x}{3}}\right) + (-\cos x)\left(9e^{\frac{x}{3}}\right)\right] + C$$

Optionally, you can clean up this answer a bit by factoring:

$$\int e^{\frac{x}{3}}\sin x\, dx = \frac{3}{10}e^{\frac{x}{3}}(\sin x - 3\cos x) + C$$

If you're skeptical that this method really gives you the right answer, check it by differentiating by using the Product Rule:

$$\frac{d}{dx}\left(\frac{3}{10}e^{\frac{x}{3}}(\sin x - 3\cos x) + C\right)$$

$$=\frac{3}{10}\left[\frac{d}{dx}e^{\frac{x}{3}}(\sin x - 3\cos x) + \frac{d}{dx}(\sin x - 3\cos x)\left(e^{\frac{x}{3}}\right)\right]$$

$$=\frac{3}{10}\left[\left(\frac{1}{3}e^{\frac{x}{3}}\right)(\sin x - 3\cos x) + (\cos x + 3\sin x)\left(e^{\frac{x}{3}}\right)\right]$$

At this point, algebra shows that this expression is equivalent to the original function:

$$=\frac{1}{10}\left(e^{\frac{x}{3}}\right)(\sin x - 3\cos x) + \frac{3}{10}(\cos x + 3\sin x)\left(e^{\frac{x}{3}}\right)$$

$$=\frac{1}{10}e^{\frac{x}{3}}\sin x - \frac{3}{10}e^{\frac{x}{3}}\cos x + \frac{3}{10}e^{\frac{x}{3}}\cos x + \frac{9}{10}e^{\frac{x}{3}}\sin x$$

$$=\frac{1}{10}e^{\frac{x}{3}}\sin x + \frac{9}{10}e^{\frac{x}{3}}\sin x$$

$$=e^{\frac{x}{3}}\sin x$$

Chapter 7

Trig Substitution: Knowing All the (Tri)Angles

In This Chapter
▶ Memorizing the basic trig integrals
▶ Integrating powers of sines and cosines, tangents and secants, and cotangents and cosecants
▶ Understanding the three cases for using trig substitution
▶ Avoiding trig substitution when possible

Trig substitution is another technique to throw in your ever-expanding bag of integration tricks. It allows you to integrate functions that contain radicals of polynomials such as $\sqrt{4-x^2}$ and other similar difficult functions.

Trig substitution may remind you of variable substitution, which I discuss in Chapter 5. With both types of substitution, you break the function that you want to integrate into pieces and express each piece in terms of a new variable. With trig substitution, however, you express these pieces as trig functions.

So, before you can do trig substitution, you need to be able to integrate a wider variety of products and powers of trig functions. The first few parts of this chapter give you the skills that you need. After that, I show you how to use trig substitution to express very complicated-looking radical functions in terms of trig functions.

Integrating the Six Trig Functions

You already know how to integrate $\sin x$ and $\cos x$ from Chapter 4, but for completeness, here are the integrals of all six trig functions:

$$\int \sin x \, dx = -\cos x + C$$

$$\int \cos x \, dx = \sin x + C$$

$$\int \tan x \, dx = \ln|\sec x| + C$$

$$\int \cot x \, dx = \ln|\sin x| + C$$

$$\int \sec x \, dx = \ln|\sec x + \tan x| + C$$

$$\int \csc x \, dx = \ln|\csc x - \cot x| + C$$

Please commit these to memory — you need them! For practice, you can also try differentiating each result to show why each of these integrals is correct.

Integrating Powers of Sines and Cosines

Later in this chapter, when I show you trig substitution, you need to know how to integrate powers of sines and cosines in a variety of combinations. In this section, I show you what you need to know.

Odd powers of sines and cosines

You can integrate *any* function of the form $\sin^m x \cos^n x$ when m is odd, for any real value of n. For this procedure, keep in mind the handy trig identity $\sin^2 x + \cos^2 x = 1$. For example, here's how you integrate $\sin^7 x \cos^{\frac{1}{3}} x$:

1. **Peel off a sin x and place it next to the dx:**

$$\int \sin^7 x \cos^{\frac{1}{3}} x \, dx = \int \sin^6 x \cos^{\frac{1}{3}} x \sin x \, dx$$

2. **Apply the trig identity $\sin^2 x = 1 - \cos^2 x$ to express the rest of the sines in the function as cosines:**

$$= \int \left(1 - \cos^2 x\right)^3 \cos^{\frac{1}{3}} x \sin x \, dx$$

3. **Use the variable substitution $u = \cos x$ and $du = -\sin x \, dx$:**

$$= -\int \left(1 - u^2\right)^3 u^{\frac{1}{3}} \, du$$

Now that you have the function in terms of powers of u, the worst is over. You can expand the function out, turning it into a polynomial. This is just algebra:

$$= - \int (1 - u^2)(1 - u^2)(1 - u^2) u^{\frac{1}{3}} du$$

$$= - \int (1 - 3u^2 + 3u^4 - u^6) u^{\frac{1}{3}} du$$

$$= - \int \left(u^{\frac{1}{3}} - 3u^{\frac{7}{3}} + 3u^{\frac{13}{3}} - u^{\frac{19}{3}} \right) du$$

To continue, use the Sum Rule and Constant Multiple Rule to separate this into four integrals, as I show you in Chapter 4. Don't forget to distribute that minus sign to all four integrals!

$$= - \int u^{\frac{1}{3}} du + 3 \int u^{\frac{7}{3}} du - 3 \int u^{\frac{13}{3}} du + \int u^{\frac{19}{3}} du$$

At this point, you can evaluate each integral separately by using the Power Rule:

$$= \frac{3}{4} u^{\frac{4}{3}} + \frac{9}{10} u^{\frac{10}{3}} - \frac{9}{16} u^{\frac{16}{3}} + \frac{3}{22} u^{\frac{22}{3}} + C$$

Finally, use $u = \cos x$ to reverse the variable substitution:

$$= \frac{3}{4} \cos^{\frac{4}{3}} x + \frac{9}{10} \cos^{\frac{10}{3}} x - \frac{9}{16} \cos^{\frac{16}{3}} x + \frac{3}{22} \cos^{\frac{22}{3}} x + C$$

Notice that when you substitute back in terms of x, the power goes next to the cos rather than the x, because you're raising the entire function $\cos x$ to a power. (See Chapter 2 if you're unclear about this point.)

Similarly, you integrate *any* function of the form $\sin^m x \cos^n x$ when n is odd, for any real value of m. These steps are practically the same as those in the previous example. For example, here's how you integrate $\sin^{-4} x \cos^9 x$:

1. **Peel off a cos x and place it next to the dx:**

$$\int \sin^{-4} x \cos^9 x \, dx = \int \sin^{-4} x \cos^8 x \cos x \, dx$$

2. **Apply the trig identity $\cos^2 x = 1 - \sin^2 x$ to express the rest of the cosines in the function as sines:**

$$= \int \sin^{-4} x \left(1 - \sin^2 x \right)^4 \cos x \, dx$$

3. **Use the variable substitution $u = \sin x$ and $du = \cos x \, dx$:**

$$= \int u^{-4} (1 - u^2)^4 du$$

At this point, you can distribute the function to turn it into a polynomial and then integrate it as I show you in the previous example.

Even powers of sines and cosines

To integrate $\sin^2 x$ and $\cos^2 x$, use the two half-angle trig identities that I show you in Chapter 2:

$$\sin^2 x = \frac{1 - \cos 2x}{2}$$

$$\cos^2 x = \frac{1 + \cos 2x}{2}$$

For example, here's how you integrate $\cos^2 x$:

1. **Use the half-angle identity for cosine to rewrite the integral in terms of cos 2x:**

$$\int \cos^2 x \, dx = \int \frac{1 + \cos 2x}{2} \, dx$$

2. **Use the Constant Multiple Rule to move the denominator outside the integral:**

$$= \frac{1}{2} \int (1 + \cos 2x) \, dx$$

3. **Distribute the function and use the Sum Rule to split it into several integrals:**

$$= \frac{1}{2} \left(\int 1 \, dx + \int \cos 2x \, dx \right)$$

4. **Evaluate the two integrals separately:**

$$= \frac{1}{2} \left(x + \frac{1}{2} \sin 2x \right) + C$$

$$= \frac{1}{2} x + \frac{1}{4} \sin 2x + C$$

As a second example, here's how you integrate $\sin^2 x \cos^4 x$:

1. **Use the two half-angle identities to rewrite the integral in terms of cos 2x:**

$$\int \sin^2 x \cos^4 x \, dx = \int \frac{1 - \cos 2x}{2} \left(\frac{1 + \cos 2x}{2} \right)^2 dx$$

2. **Use the Constant Multiple Rule to move the denominators outside the integral:**

$$= \frac{1}{8} \int (1 - \cos 2x)(1 + \cos 2x)^2 \, dx$$

3. **Distribute the function and use the Sum Rule to split it into several integrals:**

$$= \frac{1}{8} \left(\int 1 \, dx + \int \cos 2x \, dx - \int \cos^2 2x \, dx - \int \cos^3 2x \, dx \right)$$

4. **Evaluate the resulting odd-powered integrals by using the procedure from the earlier section "Odd powers of sines and cosines," and evaluate the even-powered integrals by returning to Step 1 of the previous example.**

Integrating Powers of Tangents and Secants

When you're integrating powers of tangents and secants, here's the rule to remember: *Eeeven* powers of *seeecants* are *eeeasy*. The threee Es in the keeey words should help you remember this rule. By the way, odd powers of tangents are also easy. You're on your own remembering this fact!

In this section, I show you how to integrate $\tan^m x \sec^n x$ for all positive integer values of m and n. You use this skill later in this chapter, when I show you how to do trig substitution.

Even powers of secants with tangents

To integrate $\tan^m x \sec^n x$ when n is even — for example, $\tan^8 x \sec^6 x$ — follow these steps:

1. **Peel off a $\sec^2 x$ and place it next to the *dx*:**

$$\int \tan^8 x \sec^6 x \, dx = \int \tan^8 x \sec^4 x \sec^2 x \, dx$$

2. **Use the trig identity $1 + \tan^2 x = \sec^2 x$ to express the remaining secant factors in terms of tangents:**

$$= \int \tan^8 x \left(1 + \tan^2 x\right)^2 \sec^2 x \, dx$$

3. **Use the variable substitution $u = \tan x$ and $du = \sec^2 x \, dx$:**

$$\int u^8 \left(1 + u^2\right)^2 du$$

At this point, the integral is a polynomial, and you can evaluate it as I show you in Chapter 4.

Odd powers of tangents with secants

To integrate $\tan^m x \sec^n x$ when m is odd — for example, $\tan^7 x \sec^9 x$ — follow these steps:

1. **Peel off a tan x and a sec x and place them next to the dx:**

$$\int \tan^7 x \sec^9 x = \int \tan^6 x \sec^8 x \sec x \tan x \, dx$$

2. **Use the trig identity $\tan^2 x = \sec^2 x - 1$ to express the remaining tangent factors in terms of secants:**

$$= \int \left(\sec^2 x - 1 \right)^3 \sec^8 x \sec x \tan x \, dx$$

3. **Use the variable substitution $u = \sec x$ and $du = \sec x \tan x \, dx$:**

$$= \int \left(u^2 - 1 \right)^3 u^8 \, du$$

At this point, the integral is a polynomial, and you can evaluate it as I show you in Chapter 4.

Odd powers of tangents without secants

To integrate $\tan^m x$ when m is odd, use a trig identity to convert the function to sines and cosines as follows:

$$\int \tan^m x \, dx = \int \frac{\sin^m x}{\cos^m x} \, dx = \int \sin^m x \cos^{-m} x \, dx$$

After that, you can integrate by using the procedure from the earlier section, "Odd powers of sines and cosines."

Even powers of tangents without secants

To integrate $\tan^m x$ when m is even — for example, $\tan^8 x$ — follow these steps:

1. **Peel off a $\tan^2 x$ and use the trig identity $\tan^2 x = \sec^2 x - 1$ to express it in terms of tan x:**

$$\int \tan^8 x \, dx = \int \tan^6 x \left(\sec^2 x - 1 \right) dx$$

2. **Distribute to split the integral into two separate integrals:**

$$= \int \tan^6 x \sec^2 x \, dx - \int \tan^6 x \, dx$$

3. **Evaluate the first integrals using the procedure I show you in the earlier section "Even powers of secants with tangents."**

4. **Return to Step 1 to evaluate the second integral.**

Even powers of secants without tangents

To integrate $\sec^n x$ when n is even — for example, $\sec^4 x$ — follow these steps:

1. **Use the trig identity $1 + \tan^2 x = \sec^2 x$ to express the function in terms of tangents:**

$$\int \sec^4 x \, dx = \int \left(1 + \tan^2 x\right)^2 dx$$

2. **Distribute and split the integral into three or more integrals:**

$$= \int 1 \, dx + 2 \int \tan^2 x \, dx + \int \tan^4 x \, dx$$

3. **Integrate all powers of tangents by using the procedures from the sections on powers of tangents without secants.**

Odd powers of secants without tangents

This is the hardest case, so fasten your seat belt. To integrate $\sec^n x$ when n is odd — for example, $\sec^3 x$ — follow these steps:

1. **Peel off a $\sec x$:**

$$\int \sec^3 x \, dx = \int \sec^2 x \sec x \, dx$$

2. **Use the trig identity $1 + \tan^2 x = \sec^2 x$ to express the remaining secants in terms of tangents:**

$$= \int \left(1 + \tan^2 x\right) \sec x \, dx$$

3. **Distribute and split the integral into two or more integrals:**

$$= \int \sec x + \int \tan^2 x \sec x \, dx$$

4. **Evaluate the first integral:**

$$= \ln|\sec x + \tan x| + \int \tan^2 x \sec x \, dx$$

You can omit the constant C because you still have an integral that you haven't evaluated yet — just don't forget to put it in at the end.

5. **Integrate the second integral by parts by differentiating tan x and integrating sec x tan x (see Chapter 6 for more on integration by parts):**

$$= \ln|\sec x + \tan x| + \tan x \sec x - \int \sec^3 x \, dx$$

At this point, notice that you've shown the following equation to be true:

$$\int \sec^3 x \, dx = \ln|\sec x + \tan x| + \tan x \sec x - \int \sec^3 x \, dx$$

6. **Follow the algebraic procedure that I outline in Chapter 6.**

First, substitute the variable I for the integral on both sides of the equation:

$$I = \ln|\sec x + \tan x| + \tan x \sec x - I$$

Now, solve this equation for I:

$$2I = \ln|\sec x + \tan x| + \tan x \sec x$$

$$I = \frac{1}{2} \ln|\sec x + \tan x| + \frac{1}{2} \tan x \sec x$$

Now, you can substitute the integral back for I. Don't forget, however, that you need to add a constant to the right side of this equation, to cover all possible solutions to the integral:

$$\int \sec^3 x \, dx = \frac{1}{2} \ln|\sec x + \tan x| + \frac{1}{2} \tan x \sec x + C$$

That's your final answer. I truly hope that you never have to integrate $\sec^5 x$, let alone higher odd powers of a secant. But if you do, the basic procedure I outline here will provide you with a value for $\int \sec^5 x \, dx$ in terms of $\int \sec^3 x \, dx$. Good luck!

Even powers of tangents with odd powers of secants

To integrate $\tan^m x \sec^n x$ when m is even and n is odd, transform the function into an odd power of secants, and then use the method that I outline in the previous section "Odd powers of secants without tangents."

For example, here's how you integrate $\tan^4 x \sec^3 x$:

1. **Use the trusty trig identity $\tan^2 x = \sec^2 x - 1$ to convert all the tangents to secants:**

$$\int \tan^4 x \sec^3 x \, dx = \int (\sec^2 x - 1)^2 \sec^3 x \, dx$$

2. **Distribute the function and split the integral by using the Sum Rule:**

$$= \int \sec^7 x \, dx - \int 2 \sec^5 x \, dx + \int \sec^3 x \, dx$$

3. **Solve the resulting odd-powered integrals by using the procedure from "Odd powers of secants without tangents."**

Unfortunately, this procedure brings you back to the most difficult case in this section. Fortunately, most teachers are fairly merciful when you're working with these functions, so you probably won't have to face this integral on an exam. If you do, however, you have my deepest sympathy.

Integrating Powers of Cotangents and Cosecants

The methods for integrating powers of cotangents and cosecants are very close to those for tangents and secants, which I show you in the preceding section. For example, in the earlier section "Even powers of secants with tangents," I show you how to integrate $\tan^8 x \sec^6 x$. Here's how to integrate $\cot^8 x \csc^6 x$:

1. **Peel off a $\csc^2 x$ and place it next to the dx:**

$$\int \cot^8 x \csc^6 x \, dx = \int \cot^8 x \csc^4 x \csc^2 x \, dx$$

2. **Use the trig identity $1 + \cot^2 x = \csc^2 x$ to express the remaining cosecant factors in terms of cotangents:**

$$= \int \cot^8 x \left(1 + \cot^2 x\right)^2 \csc^2 x \, dx$$

3. **Use the variable substitution $u = \cot x$ and $du = -\csc^2 x \, dx$:**

$$= -\int u^8 \left(1 + u^2\right)^2 du$$

At this point, the integral is a polynomial, and you can evaluate it as I show you in Chapter 4.

Notice that the steps here are virtually identical to those for tangents and secants. The biggest change here is the introduction of a minus sign in Step 3. So, to find out everything you need to know about integrating cotangents and cosecants, try all the examples in the previous section, but switch every tangent to a cotangent and every secant to a cosecant.

Sometimes, knowing how to integrate cotangents and cosecants can be useful for integrating negative powers of other trig functions — that is, powers of trig functions in the denominator of a fraction.

For example, suppose that you want to integrate $\dfrac{\cos^2 x}{-\sin^6 x}$. The methods that I outline earlier don't work very well in this case, but you can use trig identities to express it as cotangents and cosecants.

$$\frac{\cos^2 x}{\sin^6 x} = \frac{\cos^2 x}{\sin^2 x} \cdot \frac{1}{\sin^4 x} = \cot^2 x \csc^4 x$$

I show you more about this in the next section "Integrating Weird Combinations of Trig Functions."

Integrating Weird Combinations of Trig Functions

You don't really have to know how to integrate *every* possible trig function to pass Calculus II. If you can do all the techniques that I introduce earlier in this chapter — and I admit that's a lot to ask! — then you'll be able to handle most of what your professor throws at you with ease. You'll also have a good shot at hitting any curveballs that come at you on an exam.

But in case you're nervous about the exam and would rather study than worry, in this section I show you how to integrate a wider variety of trig functions. I don't promise to cover *all* possible trig functions exhaustively. But I do give you a few additional ways to think about and categorize trig functions that could help you when you're in unfamiliar territory.

Using identities to tweak functions

You can express every product of powers of trig functions, no matter how weird, as the product of any pair of trig functions. The three most useful pairings (as you may guess from earlier in this chapter) are sine and cosine, tangent and secant, and cotangent and cosecant. Table 7-1 shows you how to express all six trig functions as each of these pairings.

Table 7-1	Expressing the Six Trig Functions As a Pair of Trig Functions		
Trig Function	*As Sines & Cosines*	*As Tangents & Secants*	*As Cotangents & Cosecants*
$\sin x$	$\sin x$	$\dfrac{\tan x}{\sec x}$	$\dfrac{1}{\csc x}$
$\cos x$	$\cos x$	$\dfrac{1}{\sec x}$	$\dfrac{\cot x}{\csc x}$
$\tan x$	$\dfrac{\sin x}{\cos x}$	$\tan x$	$\dfrac{1}{\cot x}$
$\cot x$	$\dfrac{\cos x}{\sin x}$	$\dfrac{1}{\tan x}$	$\cot x$
$\sec x$	$\dfrac{1}{\cos x}$	$\sec x$	$\dfrac{\csc x}{\cot x}$
$\csc x$	$\dfrac{1}{\sin x}$	$\dfrac{\sec x}{\tan x}$	$\csc x$

For example, look at the following function:

$$\frac{\cos x \cot^3 x \csc^2 x}{\sin^2 x \tan x \sec x}$$

As it stands, you can't do much to integrate this monster. But try expressing it in terms of each possible pairing of trig functions:

$$= \frac{\cos^6 x}{\sin^8 x}$$

$$= \frac{\sec^2 x}{\tan^8 x}$$

$$= \cot^6 x \csc^2 x$$

As it turns out, the most useful pairing for integration in this case is $\cot^6 x$ $\csc^2 x$. No fraction is present — that is, both terms are raised to positive powers — and the cosecant term is raised to an even power, so you can use the same basic procedure that I show you in the earlier section "Even powers of secants with tangents."

Using Trig Substitution

Trig substitution is similar to variable substitution (which I discuss in Chapter 5), using a change in variable to turn a function that you can't integrate into one that you can. With variable substitution, you typically use the variable u. With trig substitution, however, you typically use the variable θ.

Trig substitution allows you to integrate a whole slew of functions that you can't integrate otherwise. These functions have a special, uniquely scary look about them, and are variations on these three themes:

$$(a^2 - bx^2)^n$$

$$(a^2 + bx^2)^n$$

$$(bx^2 - a^2)^n$$

Trig substitution is most useful when n is $\frac{1}{2}$ or a negative number — that is, for hairy square roots and polynomials in the denominator of a fraction. When n is a positive integer, your best bet is to express the function as a polynomial and integrate it as I show you in Chapter 4.

In this section, I show you how to use trig substitution to integrate functions like these. But, before you begin, take this simple test:

Trig substitution is:

✔ A) Easy and *fun* — even a child can do it!

✔ B) Not so bad when you know how.

✔ C) About as attractive as drinking bleach.

I wish I could tell you that the answer is A, but then I'd be a big liarmouth and you'd never trust me again. So I admit that trig substitution is less fun than a toga party with a hot date. At the same time, your worst trig substitution nightmares don't have to come true, so please put the bottle of bleach back in the laundry room.

I have the system right here, and if you follow along closely, I give you the tool that you need to make trig substitution mostly a matter of filling in the blanks. Trust me — have I ever lied to you?

Distinguishing three cases for trig substitution

Trig substitution is useful for integrating functions that contain three very recognizable types of polynomials in either the numerator or denominator. Table 7-2 lists the three cases that you need to know about.

Table 7-2	The Three Trig Substitution Cases	
Case	*Radical of Polynomial*	*Example*
Sine case	$(a^2 - bx^2)^n$	$\int \sqrt{4 - x^2} \, dx$
Tangent case	$(a^2 + bx^2)^n$	$\int \dfrac{1}{\left(4 + 9x^2\right)^2} \, dx$
Secant case	$(bx^2 - a^2)^n$	$\int \dfrac{1}{\sqrt{16x^2 - 1}} \, dx$

The first step to trig substitution is being able to recognize and distinguish these three cases when you see them.

Knowing the formulas for differentiating the inverse trig functions can help you remember these cases.

$$\frac{d}{dx} \arcsin x = \frac{1}{\sqrt{1 - x^2}}$$

$$\frac{d}{dx} \arctan x = \frac{1}{1 + x^2}$$

$$\frac{d}{dx} \operatorname{arcsec} x = \frac{1}{x \sqrt{x^2 - 1}}$$

Note that the differentiation formula for arcsin x contains a polynomial that looks like the sine case: a constant minus x^2. The formula for arctan x contains a polynomial that looks like the tangent case: a constant plus x^2. And the formula for arcsec x contains a polynomial that looks like the secant case: x^2 minus a constant. So, if you already know these formulas, you don't have to memorize any additional information.

Integrating the three cases

Trig substitution is a five-step process:

1. **Draw the trig substitution triangle for the correct case.**

2. **Identify the separate pieces of the integral (including dx) that you need to express in terms of θ.**

3. **Express these pieces in terms of trig functions of θ.**

4. **Rewrite the integral in terms of θ and evaluate it.**

5. **Substitute x for θ in the result.**

Don't worry if these steps don't make much sense yet. In this section, I show you how to do trig substitution for each of the three cases.

The sine case

When the function you're integrating includes a term of the form $(a^2 - bx^2)^n$, draw your trig substitution triangle for the *sine case*. For example, suppose that you want to evaluate the following integral:

$$\int \sqrt{4 - x^2}\, dx$$

This is a sine case, because a constant minus a multiple of x^2 is being raised to a power $\left(\frac{1}{2}\right)$. Here's how you use trig substitution to handle the job:

1. **Draw the trig substitution triangle for the correct case.**

   Figure 7-1 shows you how to fill in the triangle for the sine case. Notice that the radical goes on the *adjacent* side of the triangle. Then, to fill in the other two sides of the triangle, I use the square roots of the two terms inside the radical — that is, 2 and x. I place 2 on the hypotenuse and x on the opposite side.

   You can check to make sure that this placement is correct by using the Pythagorean theorem: $x^2 + \left(\sqrt{4 - x^2}\right)^2 = 2^2$.

Figure 7-1:
A trig substitution triangle for the sine case.

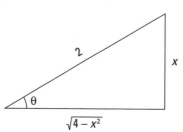

2. **Identify the separate pieces of the integral (including dx) that you need to express in terms of θ.**

   In this case, the function contains two separate pieces that contain x: $\sqrt{4 - x^2}$ and dx.

3. **Express these pieces in terms of trig functions of θ.**

   This is the real work of trig substitution, but when your triangle is set up properly, this work becomes a lot easier. In the sine case, *all* trig functions should be sines and cosines.

To represent the radical portion as a trig function of θ, first build a fraction using the radical $\sqrt{4-x^2}$ as the numerator and the constant 2 as the denominator. Then set this fraction equal to the appropriate trig function:

$$\frac{\sqrt{4-x^2}}{2} = \cos\theta$$

Because the numerator is the adjacent side of the triangle and the denominator is the hypotenuse $\left(\frac{A}{H}\right)$, this fraction is equal to ccs θ. Now, a little algebra gets the radical alone on one side of the equation:

$$\sqrt{4-x^2} = 2\cos\theta$$

Next, you want to express dx as a trig function of θ. To do so, build another fraction with the variable x in the numerator and the constant 2 in the denominator. Then set this fraction equal to the correct trig function:

$$\frac{x}{2} = \sin\theta$$

This time, the numerator is the opposite side of the triangle and the denominator is the hypotenuse $\left(\frac{O}{H}\right)$, so this fraction is equal to sin θ. Now, solve for x and then differentiate:

$$x = 2\sin\theta$$

$$dx = 2\cos\theta\,d\theta$$

4. **Rewrite the integral in terms of θ and evaluate it:**

$$\int \sqrt{4-x^2}\,dx$$

$$\int 2\cos\theta \cdot 2\cos\theta\,d\theta$$

$$= 4\int \cos^2\theta\,d\theta$$

Knowing how to evaluate trig integrals really pays off here. I cut to the chase in this example, but earlier in this chapter (in "Integrating Powers of Sines and Cosines"), I show you how to integrate all sorts of trig functions like this one:

$$= 2\theta + \sin 2\theta + C$$

5. **To change those two θ terms into x terms, reuse the following equation:**

$$\frac{x}{2} = \sin\theta$$

$$\theta = \arcsin\frac{x}{2}$$

So here's a substitution that gives you an answer:

$$= 2 \arcsin \frac{x}{2} + \sin(2 \arcsin \frac{x}{2}) + C$$

This answer is perfectly valid so, technically speaking, you can stop here. However, some professors frown upon the nesting of trig and inverse trig functions, so they'll prefer a simplified version of $\sin(2 \arcsin \frac{x}{2})$. To find this, start by applying the double-angle sine formula (see Chapter 2) to $\sin 2\theta$:

$$\sin 2\theta = 2 \sin \theta \cos \theta$$

Now, use your trig substitution triangle to substitute values for $\sin \theta$ and $\cos \theta$ in terms of x:

$$= 2 \left(\frac{x}{2} \right) \left(\frac{\sqrt{4 - x^2}}{2} \right)$$

$$= \frac{1}{2} x \sqrt{4 - x^2}$$

To finish up, substitute this expression for that problematic second term to get your final answer in a simplified form:

$$2\theta + \sin 2\theta + C$$

$$= 2 \arcsin \frac{x}{2} + \frac{1}{2} x \sqrt{4 - x^2} + C$$

The tangent case

When the function you're integrating includes a term of the form $(a^2 + x^2)^n$, draw your trig substitution triangle for the *tangent case*. For example, suppose that you want to evaluate the following integral:

$$\int \frac{1}{\left(4 + 9x^2 \right)^2} \, dx$$

This is a tangent case, because a constant plus a multiple of x^2 is being raised to a power (-2). Here's how you use trig substitution to integrate:

1. **Draw the trig substitution triangle for the tangent case.**

   Figure 7-2 shows you how to fill in the triangle for the tangent case. Notice that the radical of what's inside the parentheses goes on the *hypotenuse* of the triangle. Then, to fill in the other two sides of the triangle, use the square roots of the two terms inside the radical — that is, 2 and $3x$. Place the constant term 2 on the adjacent side and the variable term $3x$ on the opposite side.

With the tangent case, make sure not to mix up your placement of the variable and the constant.

Figure 7-2:
A trig
substitution
triangle for
the tangent
case.

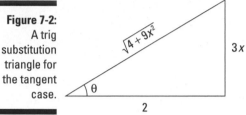

2. **Identify the separate pieces of the integral (including dx) that you need to express in terms of θ.**

   In this case, the function contains two separate pieces that contain x:

   $\dfrac{1}{\left(4+9x^2\right)^2}$ and dx.

3. **Express these pieces in terms of trig functions of θ.**

   In the tangent case, *all* trig functions should be initially expressed as tangents and secants.

   To represent the rational portion as a trig function of θ, build a fraction using the radical $\sqrt{4+9x^2}$ as the numerator and the constant 2 as the denominator. Then set this fraction equal to the appropriate trig function:

   $$\dfrac{\sqrt{4+9x^2}}{2} = \sec\theta$$

   Because this fraction is the hypotenuse of the triangle over the adjacent side $\left(\dfrac{H}{A}\right)$, it's equal to sec θ. Now, use algebra and trig identities to tweak this equation into shape:

   $$\sqrt{4+9x^2} = 2\sec\theta$$
   $$\left(4+9x^2\right)^2 = 8\sec^4\theta$$
   $$\dfrac{1}{\left(4+9x^2\right)^2} = \dfrac{1}{8\sec^4\theta}$$

   Next, express dx as a trig function of θ. To do so, build another fraction with the variable $3x$ in the numerator and the constant 2 in the denominator:

   $$\dfrac{3x}{2} = \tan\theta$$

This time, the fraction is the opposite side of the triangle over the adjacent side $\left(\dfrac{O}{A}\right)$, so it equals $\tan\theta$. Now, solve for x and then differentiate:

$$x = \frac{2}{3}\tan\theta$$
$$dx = \frac{2}{3}\sec^2\theta\,d\theta$$

4. **Express the integral in terms of θ and evaluate it:**

$$\int \frac{1}{\left(4+9x^2\right)^2}\,dx$$
$$= \int \frac{1}{8\sec^4\theta}\cdot\frac{2}{3}\sec^2\theta\,d\theta$$

Now, some cancellation and reorganization turns this nasty-looking integral into something manageable:

$$= \frac{1}{12}\int \cos^2\theta\,d\theta$$

At this point, use your skills from the earlier section "Even Powers of Sines and Cosines" to evaluate this integral:

$$= \frac{1}{24}\theta + \frac{1}{48}\sin 2\theta + C$$

5. **Change the two θ terms back into x terms:**

You need to find a way to express θ in terms of x. Here's the simplest way:

$$\tan\theta = \frac{3x}{2}$$
$$\theta = \arctan\frac{3x}{2}$$

So here's a substitution that gives you an answer:

$$\frac{1}{24}\theta + \frac{1}{48}\sin 2\theta + C = \frac{1}{24}\arctan\frac{3x}{2} + \frac{1}{48}\sin\left(2\arctan\frac{3x}{2}\right) + C$$

This answer is valid, but most professors won't be crazy about that ugly second term, with the sine of an arctangent. To simplify it, apply the double-angle sine formula (see Chapter 2) to $\frac{1}{48}\sin 2\theta$:

$$\frac{1}{48}\sin 2\theta = \frac{1}{24}\sin\theta\cos\theta$$

Now, use your trig substitution triangle to substitute values for $\sin\theta$ and $\cos\theta$ in terms of x:

$$= \frac{1}{24}\left(\frac{3x}{\sqrt{4+9x^2}}\right)\left(\frac{2}{\sqrt{4+9x^2}}\right)$$

$$= \frac{6x}{24\left(4+9x^2\right)}$$

$$= \frac{x}{\left(16+36x^2\right)}$$

Finally, use this result to express the answer in terms of x:

$$\frac{1}{24}\theta + \frac{1}{48}\sin 2\theta + C$$

$$= \frac{1}{24}\arctan\frac{3x}{2} + \frac{x}{\left(16+36x^2\right)} + C$$

The secant case

When the function that you're integrating includes a term of the form $(bx^2 - a^2)^n$, draw your trig substitution triangle for the *secant case*. For example, suppose that you want to evaluate this integral:

$$\int \frac{1}{\sqrt{16x^2-1}}\,dx$$

This is a secant case, because a multiple of x^2 minus a constant is being raised to a power $\left(-\frac{1}{2}\right)$. Integrate by using trig substitution as follows:

1. **Draw the trig substitution triangle for the secant case.**

   Figure 7-3 shows you how to fill in the triangle for the secant case. Notice that the radical goes on the *opposite* side of the triangle. Then, to fill in the other two sides of the triangle, use the square roots of the two terms inside the radical — that is, 1 and $4x$. Place the constant 1 on the adjacent side and the variable $4x$ on the hypotenuse.

   You can check to make sure that this placement is correct by using the Pythagorean theorem: $1^2 + \left(\sqrt{16x^2-1}\right)^2 = (4x)^2$.

Figure 7-3:
A trig
substitution
triangle for
the secant
case.

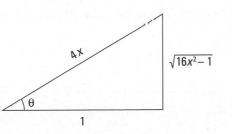

2. **Identify the separate pieces of the integral (including *dx*) that you need to express in terms of θ.**

   In this case, the function contains two separate pieces that contain *x*:

   $$\frac{1}{\sqrt{16x^2 - 1}} \text{ and } dx.$$

3. **Express these pieces in terms of trig functions of θ.**

   In the secant case (as in the tangent case), *all* trig functions should be initially represented as tangents and secants.

   To represent the radical portion as a trig function of θ, build a fraction by using the radical $\sqrt{16x^2 - 1}$ as the numerator and the constant 1 as the denominator. Then set this fraction equal to the appropriate trig function:

   $$\frac{\sqrt{16x^2 - 1}}{1} = \tan\theta$$

   Notice that this fraction is the opposite side of the triangle over the adjacent side $\left(\frac{O}{A}\right)$, so it equals tan θ. Simplifying it a bit gives you this equation:

   $$\frac{1}{\sqrt{16x^2 - 1}} = \frac{1}{\tan\theta}$$

   Next, express *dx* as a trig function of θ. To do so, build another fraction with the variable *x* in the numerator and the constant 1 in the denominator:

   $$\frac{4x}{1} = \sec\theta$$

   This time, the fraction is the hypotenuse over the adjacent side of the triangle $\left(\frac{H}{A}\right)$, which equals sec θ. Now, solve for *x* and differentiate to find *dx*:

   $$x = \frac{1}{4}\sec\theta$$

   $$dx = \frac{1}{4}\sec\theta\tan\theta\, d\theta$$

4. **Express the integral in terms of θ and evaluate it:**

   $$\int \frac{1}{\sqrt{16x^2 - 1}}\, dx = \int \frac{1}{\tan\theta} \cdot \frac{1}{4}\sec\theta\tan\theta\, d\theta$$

   $$= \frac{1}{4}\int \sec\theta\, d\theta$$

   Now, use the formula for the integral of the secant function from "Integrating the Six Trig Functions" earlier in this chapter:

   $$= \frac{1}{4}\ln|\sec\theta + \tan\theta| + C$$

5. Change the two θ terms back into *x* terms:

In this case, you don't have to find the value of θ because you already know the values of sec θ and tan θ in terms of *x* from Step 3. So, substitute these two values to get your final answer:

$$= \frac{1}{4} \ln \left| 4x + \sqrt{16x^2 - 1} \right| + C$$

Knowing when to avoid trig substitution

Now that you know how to use trig substitution, I give you a skill that can be even more valuable: *avoiding* trig substitution when you don't need it. For example, look at the following integral:

$$\int \left(1 - 4x^2 \right)^2 dx$$

This might look like a good place to use trig substitution, but it's an even better place to use a little algebra to expand the problem into a polynomial:

$$= \int \left(1 - 8x^2 + 16x^4 \right) dx$$

Similarly, look at this integral:

$$\int \frac{x}{\sqrt{x^2 - 49}} dx$$

You can use trig substitution to evaluate this integral if you want to. (You can also walk to the top of the Empire State Building instead of taking the elevator if that tickles your fancy.) However, the presence of that little *x* in the numerator should tip you off that variable substitution will work just as well (flip to Chapter 5 for more on variable substitution):

Let $u = x^2 - 49$

$du = 2x\, dx$

$\frac{1}{2} du = x\, dx$

Using this substitution results in the following integral:

$$= \frac{1}{2} \int \frac{1}{\sqrt{u}}\, du$$

$$= \sqrt{u} + C$$

$$= \sqrt{x^2 - 49} + C$$

Done! I probably don't need to tell you how much time and aggravation you can save by working smarter rather than harder. So I won't!

Chapter 8

When All Else Fails: Integration with Partial Fractions

....

....

*L*et's face it: At this point in your math career, you have bigger things to worry about than adding a couple of fractions. And if you've survived integration by parts (Chapter 6) and trig integration (Chapter 7), multiplying a few polynomials isn't going to kill you either.

So, here's the good news about partial fractions: They're based on very simple arithmetic and algebra. In this chapter, I introduce you to the basics of partial fractions and show you how to use them to evaluate integrals. I illustrate four separate cases in which partial fractions can help you integrate functions that would otherwise be a big ol' mess.

Now, here's the bad news: Although the concept of partial fractions isn't difficult, using them to integrate is just about the most tedious thing you encounter in this book. And as if that weren't enough, partial fractions only work with *proper* rational functions, so I show you how to distinguish these from their ornery cousins, *improper* rational functions. I also give you a big blast from the past, a refresher on *polynomial division,* which I promise is easier than you remember it to be.

Strange but True: Understanding Partial Fractions

Partial fractions are useful for integrating *rational functions* — that is, functions in which a polynomial is divided by a polynomial. The basic tactic behind partial fractions is to split up a rational function that you can't integrate into two or more simpler functions that you can integrate.

In this section, I show you a simple analogy for partial fractions that involves only arithmetic. After you understand this analogy, partial fractions make a lot more sense. At the end of the section, I show you how to solve an integral by using partial fractions.

Looking at partial fractions

Suppose that you want to split the fraction $\frac{14}{15}$ into a sum of two smaller fractions. Start by decomposing the denominator down to its factors — 3 and 5 — and setting the denominators of these two smaller fractions to these numbers:

$$\frac{14}{15} = \frac{A}{3} + \frac{B}{5} = \frac{5A + 3B}{15}$$

So, you want to find an A and a B that satisfy this equation:

$$5A + 3B = 14$$

Now, just by eyeballing this fraction, you can probably find the nice integer solution $A = 1$ and $B = 3$, so:

$$\frac{14}{15} = \frac{1}{3} + \frac{3}{5}$$

If you include negative fractions, you can find integer solutions like this for every fraction. For example, the fraction $\frac{1}{15}$ seems too small to be a sum of thirds and fifths, until you discover:

$$\frac{2}{3} - \frac{3}{5} = \frac{1}{15}$$

Using partial fractions with rational expressions

The technique of breaking up fractions works for rational expressions. It can provide a strategy for integrating functions that you can't compute directly. For example, suppose that you're trying to find this integral:

$$\int \frac{6}{x^2 - 9}\, dx$$

You can't integrate this function directly, but if you break it into the sum of two simpler rational expressions, you can use the Sum Rule to solve them separately. And, fortunately, the polynomial in the denominator factors easily:

$$\frac{6}{x^2 - 9} = \frac{6}{(x+3)(x-3)}$$

So, set up this polynomial fraction just as I do with the regular fractions in the previous section:

$$\frac{6}{(x+3)(x-3)} = \frac{A}{x+3} + \frac{B}{x-3}$$
$$= \frac{A(x-3) + B(x+3)}{(x+3)(x-3)}$$

This gives you the following equation:

$$A(x-3) + B(x+3) = 6$$

This equation works for *all* values of x. You can exploit this fact to find the values of A and B by picking helpful values of x. To solve this equation for A and B, substitute the roots of the original polynomial (3 and –3) for x and watch what happens:

$A(3-3) + B(3+3) = 6$	$A(-3-3) + B(-3+3) = 6$
$6B = 6$	$-6A = 6$
$B = 1$	$A = -1$

Now substitute these values of A and B back into the rational expressions:

$$\frac{6}{(x+3)(x-3)} = -\frac{1}{x+3} + \frac{1}{x-3}$$

This sum of two rational expressions is a whole lot friendlier to integrate than what you started with. Use the Sum Rule followed by a simple variable substitution (see Chapter 5):

$$\int \left(-\frac{1}{x+3} + \frac{1}{x-3} \right) dx$$

$$= -\int \frac{1}{x+3}\, dx + \int \frac{1}{x-3}\, dx$$

$$= -\ln |x+3| + \ln |x-3| + C$$

As with regular fractions, you can't always break rational expressions apart in this fashion. But in four distinct cases, which I discuss in the next section, you can use this technique to integrate complicated rational functions.

Solving Integrals by Using Partial Fractions

In the last section, I show you how to use partial fractions to split a complicated rational function into several smaller and more-manageable functions. Although this technique will certainly amaze your friends, you may be wondering why it's worth learning.

The payoff comes when you start integrating. Lots of times, you can integrate a big rational function by breaking it into the sum of several bite-sized chunks. Here's a bird's-eye view of how to use partial fractions to integrate a rational expression:

1. **Set up the rational expression as a sum of partial fractions with unknowns (A, B, C, and so forth) in the numerators.**

    I call these *unknowns* rather than variables to distinguish them from x, which remains a variable for the whole problem.

2. **Find the values of all the unknowns and plug them into the partial fractions.**

3. **Integrate the partial fractions separately by whatever method works.**

In this section, I focus on these three steps. I show you how to turn a complicated rational function into a sum of simpler rational functions and how to replace unknowns (such as A, B, C, and so on) with numbers. Finally, I give you a few important techniques for integrating the types of simpler rational functions that you often see when you use partial fractions.

Setting up partial fractions case by case

Setting up a sum of partial fractions isn't difficult, but there are four distinct cases to watch out for. Each case results in a different setup — some easier than others.

Try to get familiar with these four cases, because I use them throughout this chapter. Your first step in any problem that involves partial fractions is to recognize which case you're dealing with so that you can solve the problem.

Each of these cases is listed in Table 8-1.

Table 8-1	The Four Cases for Setting up Partial Fractions	
Case	**Example**	**As Partial Fractions**
Case #1: Distinct linear factors	$\dfrac{x}{(x+4)(x-7)}$	$\dfrac{A}{x+4} + \dfrac{B}{x-7}$
Case #2: Distinct irreducible quadratic factors	$\dfrac{8}{(x^2+3)(x^2+9)}$	$\dfrac{A+Bx}{\left(x-\sqrt{3}\right)\left(x+\sqrt{3}\right)} + \dfrac{C+Dx}{x^2+9}$
Case #3: Repeated linear factors	$\dfrac{2x+2}{(x+5)^2}$	$\dfrac{A}{x+5} + \dfrac{B}{(x+5)^2}$
Case #4: Repeated quadratic factors	$\dfrac{x^2-2}{\left(x^2+6\right)^2}$	$\dfrac{A+Bx}{x^2+6} + \dfrac{C+Dx}{\left(x^2+6\right)^2}$

Case #1: Distinct linear factors

The simplest case in which partial fractions are helpful is when the denominator is the product of *distinct linear factors* — that is, linear factors that are nonrepeating.

For each distinct linear factor in the denominator, add a partial fraction of the following form:

$$\frac{A}{\text{linear factor}}$$

For example, suppose that you want to integrate the following rational expression:

$$\frac{1}{x(x+2)(x-5)}$$

The denominator is the product of three distinct linear factors — x, $(x + 2)$, and $(x - 5)$ — so it's equal to the sum of three fractions with these factors as denominators:

$$= \frac{A}{x} + \frac{B}{x+2} + \frac{C}{x-5}$$

The number of distinct linear factors in the denominator of the original expression determines the number of partial fractions. In this example, the presence of three factors in the denominator of the original expression yields three partial fractions.

Case #2: Distinct quadratic factors

Another not-so-bad case where you can use partial fractions is when the denominator is the product of *distinct quadratic factors* — that is, quadratic factors that are nonrepeating.

For each distinct quadratic factor in the denominator, add a partial fraction of the following form:

$$\frac{A + Bx}{\text{quadratic factor}}$$

For example, suppose that you want to integrate this function:

$$\frac{5x - 6}{(x - 2)(x^2 + 3)}$$

The first factor in the denominator is linear, but the second is quadratic and can't be decomposed to linear factors. So, set up your partial fractions as follows:

$$= \frac{A}{x - 2} + \frac{Bx + C}{x^2 + 3}$$

As with distinct linear factors, the number of distinct quadratic factors in the denominator tells you how many partial fractions you get. So in this example, two factors in the denominator yield two partial fractions.

Case #3: Repeated linear factors

Repeated linear factors are more difficult to work with because each factor requires more than one partial fraction.

For each squared linear factor in the denominator, add *two* partial fractions in the following form:

$$\frac{A}{\text{linear factor}} + \frac{B}{(\text{linear factor})^2}$$

For each quadratic factor in the denominator that's raised to the third power, add *three* partial fractions in the following form:

$$\frac{A}{\text{linear factor}} + \frac{B}{(\text{linear factor})^2} + \frac{C}{(\text{linear factor})^3}$$

Generally speaking, when a linear factor is raised to the *n*th power, add *n* partial fractions. For example, suppose that you want to integrate the following expression:

$$\frac{x^2 - 3}{(x+5)(x-1)^3}$$

This expression contains all linear factors, but one of these factors $(x + 5)$ is nonrepeating and the other $(x - 1)$ is raised to the third power. Set up your partial fractions this way:

$$= \frac{A}{x+5} + \frac{B}{x-1} + \frac{C}{(x-1)^2} + \frac{D}{(x-1)^3}$$

As you can see, I add one partial fraction to account for the nonrepeating factor and three to account for the repeating factor.

Case #4: Repeated quadratic factors

Your worst nightmare when it comes to partial fractions is when the denominator includes repeated quadratic factors.

For each squared quadratic factor in the denominator, add *two* partial fractions in the following form:

$$\frac{Ax + B}{\text{quadratic factor}} + \frac{Cx + D}{(\text{quadratic factor})^2}$$

For each quadratic factor in the denominator that's raised to the third power, add *three* partial fractions in the following form:

$$\frac{Ax + B}{\text{quadratic factor}} + \frac{Cx + D}{(\text{quadratic factor})^2} + \frac{Ex + F}{(\text{quadratic factor})^3}$$

Generally speaking, when a quadratic factor is raised to the *n*th power, add *n* partial fractions. For example:

$$\frac{7 + x}{(x-8)(x^2+x+1)(x^2+3)^2}$$

This denominator has one nonrepeating linear factor $(x - 8)$, one nonrepeating quadratic factor $(x^2 + x - 1)$, and one quadratic expression that's squared $(x^2 + 3)$. Here's how you set up the partial fractions:

$$= \frac{A}{x-8} + x^2 + x - 1 = \left(x - \left(-1 + \sqrt{\frac{5}{2}}\right)\right) \cdot \left(x - \left(-1 - \sqrt{\frac{5}{2}}\right)\right) + \frac{D+Ex}{x^2+3} + \frac{F+Gx}{(x^2+3)^2}$$

This time, I added one partial fraction for each of the nonrepeating factors and two partial fractions for the squared factor.

Beyond the four cases: Knowing how to set up any partial fraction

At the outset, I have some great news: You'll probably never have to set up a partial fraction any more complex than the one that I show you in the previous section. So relax.

I'm aware that some students like to get this stuff on a case-by-case basis, so that's why I introduce it that way. However, other students prefer to be shown an overall pattern, so they can get the Zen math experience. If this is your path, read on. If not, feel free to skip ahead.

You can break *any* rational function into a sum of partial fractions. You just need to understand the pattern for repeated higher-degree polynomial factors in the denominator. This pattern is simplest to understand with an example. Suppose that you're working with the following rational function:

$$\frac{5x+1}{\left(7x^4+1\right)^5(x+2)^2\left(x^2+1\right)^2}$$

In this factor, the denominator includes a problematic factor that's a *fourth-degree polynomial* raised to the *fifth power*. You can't decompose this factor further, so the function falls outside the four cases I outline earlier in this chapter. Here's how you break this rational function into partial fractions:

$$=\frac{Ax^3+Bx^2+Cx+D}{7x^4+1}+$$

$$\frac{Ex^3+Fx^2+Gx+H}{\left(7x^4+1\right)^2}+$$

$$\frac{Ix^3+Jx^2+Kx+L}{\left(7x^4+1\right)^3}+$$

$$\frac{Mx^3+Nx^2+Ox+P}{\left(7x^4+1\right)^4}+$$

$$\frac{Qx^3+Rx^2+Sx+T}{\left(7x^4+1\right)^5}+$$

$$\frac{U}{x+2}+\frac{V}{\left(x+2\right)^2}+$$

$$\frac{Wx+X}{\left(x^2+1\right)}+\frac{Yx+Z}{\left(x^2+1\right)^2}$$

As you can see, I completely run out of capital letters. As you can also see, the problematic factor spawns *five partial fractions* — that is, the same number as the power it's raised to. Furthermore:

- ✔ The numerator of each of these fractions is a polynomial of *one degree less* than the denominator.

- ✔ The denominator of each of these fractions is a carbon copy of the original denominator, but in each case raised to a different power up to and including the original.

The remaining two factors in the denominator — a repeated linear (Case #3) and a repeated quadratic (Case #4) — give you the remaining four fractions, which look tiny and simple by comparison.

Clear as mud? Spend a little time with this example and the pattern should become clearer. Notice, too, that the four cases that I outline earlier in this chapter all follow this same general pattern.

You'll probably never have to work with anything as complicated as this — let alone try to integrate it! — but when you understand the pattern, you can break any rational function into partial fractions without worrying which case it is.

Knowing the ABCs of finding unknowns

You have two ways to find the unknowns in a sum of partial fractions. The easy and quick way is by using the roots of polynomials. Unfortunately, this method doesn't always find all the unknowns in a problem, though it often finds a few of them. The second way is to set up a system of equations.

Rooting out values with roots

When a sum of partial fractions has linear factors (either distinct or repeated), you can use the roots of these linear factors to find the values of unknowns. For example, in the earlier section "Case #1: Distinct linear factors," I set up the following equation:

$$\frac{1}{x(x+2)(x-5)} = \frac{A}{x} + \frac{B}{x+2} + \frac{C}{x-5}$$

To find the values of the unknowns A, B, and C, first get a common denominator on the right side of this equation (the same denominator that's on the left side):

$$\frac{1}{x(x+2)(x-5)} = \frac{A(x+2)(x-5) + Bx(x-5) + Cx(x+2)}{x(x+2)(x-5)}$$

Now, multiply both sides by this denominator:

$$1 = A(x+2)(x-5) + Bx(x-5) + Cx(x+2)$$

To find the values of A, B, and C, substitute the roots of the three factors (0, –2, and 5):

$$1 = A(2)(-5) \qquad\qquad 1 = B(-2)(-2-5) \qquad\qquad 1 = C(5)(5+2)$$

$$A = -\frac{1}{10} \qquad\qquad\qquad B = \frac{1}{14} \qquad\qquad\qquad\qquad C = \frac{1}{35}$$

Plugging these values back into the original integral gives you:

$$-\frac{1}{10x} + \frac{1}{14(x+2)} + \frac{1}{35(x-5)}$$

This expression is equivalent to what you started with, but it's much easier to integrate. To do so, use the Sum Rule to break it into three integrals, the Constant Multiple Rule to move fractional coefficients outside each integral, and variable substitution (see Chapter 5) to do the integration. Here's the answer so that you can try it out:

$$\int \left[-\frac{1}{10x} + \frac{1}{14(x+2)} + \frac{1}{35(x-5)} \right] dx$$

$$= -\frac{1}{10}\ln x + \frac{1}{14}\ln(x+2) + \frac{1}{35}\ln(x-5) + K$$

In this answer, I use K rather than C to represent the constant of integration to avoid confusion, because I already use C in the earlier partial fractions.

Working systematically with a system of equations

Setting up a system of equations is an alternative method for finding the value of unknowns when you're working with partial fractions. It's not as simple as plugging in the roots of factors (which I show you in the last section), but it's your only option when the root of a quadratic factor is imaginary.

To illustrate this method and why you need it, I use the problem that I set up in "Case #2: Distinct quadratic factors":

$$\frac{5x-6}{(x-2)(x^2+3)} = \frac{A}{x-2} + \frac{Bx+C}{x^2+3}$$

To start out, see how far you can get by plugging in the roots of equations. As I show you in "Rooting out values with roots," begin by getting a common denominator on the right side of the equation:

$$\frac{5x-6}{(x-2)(x^2+3)} = \frac{(A)(x^2+3) + (Bx+C)(x-2)}{(x-2)(x^2+3)}$$

Now, multiply the whole equation by the denominator:

$$5x - 6 = (A)(x^2 + 3) + (Bx + C)(x - 2)$$

The root of $x - 2$ is 2, so let $x = 2$ and see what you get:

$$5(2) - 6 = A(2^2 + 3)$$
$$A = \frac{4}{7}$$

Now, you can substitute $\frac{4}{7}$ for A:

$$5x - 6 = \frac{4}{7(x^2 + 3)} + (Bx + C)(x - 2)$$

Unfortunately, $x^2 + 3$ has no root in the real numbers, so you need a different approach. First, get rid of the parentheses on the right side of the equation:

$$5x - 6 = \frac{4}{7}x^2 + \frac{12}{7} + Bx^2 - 2Bx + Cx - 2C$$

Next, combine similar terms (using x as the variable by which you judge similarity). This is just algebra, so I skip a few steps here:

$$x^2\left(\frac{4}{7} + B\right) + x(-2B + C - 5) + \left(\frac{12}{7} - 2C + 6\right) = 0$$

Because this equation works for *all* values of x, I now take what appears to be a questionable step, breaking this equation into three separate equations as follows:

$$\frac{4}{7} + B = 0$$
$$-2B + C - 5 = 0$$
$$\frac{12}{7} - 2C + 6 = 0$$

At this point, a little algebra tells you that $B = -\frac{4}{7}$ and $C = \frac{27}{7}$. So you can substitute the values of A, B, and C back into the partial fractions:

$$\frac{5x - 6}{(x - 2)(x^2 + 3)} = \frac{4}{7(x - 2)} + \frac{-\frac{4}{7}x + \frac{27}{7}}{x^2 + 3}$$

You can simplify the second fraction a bit:

$$\frac{4}{7(x - 2)} + \frac{-4x + 27}{7(x^2 + 3)}$$

Integrating partial fractions

After you express a hairy rational expression as the sum of partial fractions, integrating becomes a lot easier. Generally speaking, here's the system:

1. **Split all rational terms with numerators of the form $Ax + B$ into two terms.**

2. **Use the Sum Rule to split the entire integral into many smaller integrals.**

3. **Use the Constant Multiple Rule to move coefficients outside each integral.**

4. **Evaluate each integral by whatever method works.**

Linear factors: Cases #1 and #3

When you start out with a linear factor — whether distinct (Case #1) or repeated (Case #3) — using partial fractions leaves you with an integral in the following form:

$$\int \frac{1}{(ax+b)^n}\, dx$$

Integrate all these cases by using the variable substitution $u = ax + b$ so that $du = a\, dx$ and $\frac{du}{a} = dx$. This substitution results in the following integral:

$$= \frac{1}{a} \int \frac{1}{u^n}\, du$$

Here are a few examples:

$$\int \frac{1}{3x+5}\, dx = \frac{1}{3} \ln|3x+5| + C$$

$$\int \frac{1}{(6x-1)^2}\, dx = -\frac{1}{6(6x-1)} + C$$

$$\int \frac{1}{(x+9)^3}\, dx = -\frac{1}{2(x+9)^2} + C$$

Quadratic factors of the form $(ax^2 + C)$: Cases #2 and #4

When you start out with a quadratic factor of the form $(ax^2 + C)$ — whether distinct (Case #2) or repeated (Case #4) — using partial fractions results in the following two integrals:

$$\int \frac{x}{(ax^2+C)^n}\, dx$$

$$\int \frac{1}{(ax^2+C)^n}\, dx$$

Integrate the first by using the variable substitution $u = ax^2 + C$ so that $du = ax\, dx$ and $\frac{du}{a} = x\, dx$. This substitution results in the following integral:

$$= \frac{1}{a} \int \frac{1}{u^n}\, du$$

This is the same integral that arises in the linear case that I describe in the previous section. Here are some examples:

$$\int \frac{x}{7x^2 + 1}\, dx = \frac{1}{14} \ln\left|7x^2 + 1\right| + C$$

$$\int \frac{x}{\left(x^2 + 4\right)^2}\, dx = -\frac{2}{\left(x^2 + 4\right)} + C$$

$$\int \frac{x}{\left(8x^2 - 2\right)^3}\, dx = \frac{-1}{32\left(8x^2 - 2\right)^2} + C$$

To evaluate the second integral, use the following formula:

$$\int \frac{1}{x^2 + n^2}\, dx = \frac{1}{n} \arctan\frac{x}{n} + C$$

Quadratic factors of the form $(ax^2 + bx + C)$: Cases #2 and #4

Most math teachers have at least a shred of mercy in their hearts, so they don't tend to give you problems that include this most difficult case. When you start out with a quadratic factor of the form $(ax^2 + bx + C)$ — whether distinct (Case #2) or repeated (Case #4) — using partial fractions results in the following integral:

$$\int \frac{hx + k}{\left(ax^2 + bx + C\right)^n}\, dx$$

I know, I know — that's way too many letters and not nearly enough numbers. Here's an example:

$$\int \frac{x - 5}{x^2 + 6x + 13}\, dx$$

This is about the hairiest integral you're ever going to see at the far end of a partial fraction. To evaluate it, you want to use the variable substitution $u = x^2 + 6x + 13$ so that $du = (2x + 6)\, dx$. If the numerator were $2x + 6$, you'd be in great shape. So you need to tweak the numerator a bit. First multiply it by 2 and divide the whole integral by 2:

$$= \frac{1}{2} \int \frac{2x - 10}{x^2 + 6x + 13}\, dx$$

Because you multiplied the entire integral by 1, no net change has occurred. Now, add 16 and –16 to the numerator:

$$= \frac{1}{2} \int \frac{2x + 6 - 16}{x^2 + 6x + 13} \, dx$$

This time, you add 0 to the integral, which doesn't change its value. At this point, you can split the integral in two:

$$= \frac{1}{2} \left[\int \frac{2x + 6}{x^2 + 6x + 13} \, dx - 16 \int \frac{1}{x^2 + 6x + 13} \, dx \right]$$

At this point, you can use the desired variable substitution (which I mention a few paragraphs earlier) to change the first integral as follows:

$$\int \frac{2x + 6}{x^2 + 6x + 13} \, dx = \int \frac{1}{u} \, du$$

$$= \ln |u| + C$$

$$= \ln |x^2 + 6x + 13| + C$$

To solve the second integral, complete the square in the denominator: Divide the b term (6) by 2 and square it, and then represent the C term (13) as the sum of this and whatever's left:

$$-16 \int \frac{1}{x^2 + 6x + 9 + 4} \, dx$$

Now, split the denominator into two squares:

$$= -16 \int \frac{1}{(x + 3)^2 + 2^2} \, dx$$

To evaluate this integral, use the same formula that I show you in the previous section:

$$\int \frac{1}{x^2 + n^2} \, dx = \frac{1}{n} \arctan \frac{x}{n} + C$$

So here's the final answer for the second integral:

$$-8 \arctan \frac{x + 3}{2} + C$$

Therefore, piece together the complete answer as follows:

$$\int \frac{x - 5}{x^2 + 6x + 13} \, dx$$

$$= \frac{1}{2} \left[\ln |x^2 + 6x + 13| - 8 \arctan \frac{x + 3}{2} \right] + C$$

$$= \frac{1}{2} \ln |x^2 + 6x + 13| - 4 \arctan \frac{x + 3}{2} + C$$

Integrating Improper Rationals

Integration by partial fractions works only with *proper rational expressions,* but not with *improper rational expressions.* In this section, I show you how to tell these two beasts apart. Then I show you how to use polynomial division to turn improper rationals into more acceptable forms. Finally, I walk you through an example in which you integrate an improper rational expression by using everything in this chapter.

Distinguishing proper and improper rational expressions

Telling a proper fraction from an improper one is easy: A fraction $\frac{a}{b}$ is proper if the numerator (disregarding sign) is *less* than the denominator, and improper otherwise.

With rational expressions, the idea is similar, but instead of comparing the value of the numerator and denominator, you compare their *degrees.* The degree of a polynomial is its highest power of x (flip to Chapter 2 for a refresher on polynomials).

A rational expression is proper if the degree of the numerator is less than the degree of the denominator, and improper otherwise.

For example, look at these three rational expressions:

$$\frac{x^2 + 2}{x^3}$$

$$\frac{x^5}{3x^2 - 1}$$

$$\frac{-5x^4}{3x^4 - 2}$$

In the first example, the numerator is a second-degree polynomial and the denominator is a third-degree polynomial, so the rational is *proper.* In the second example, the numerator is a fifth-degree polynomial and the denominator is a second-degree polynomial, so the expression is *improper.* In the third example, the numerator and denominator are both fourth-degree polynomials, so the rational function is *improper.*

Recalling polynomial division

Most math students learn polynomial division in Algebra II, demonstrate that they know how to do it on their final exam, and then promptly forget it. And, happily, they never need it again — except to pass the time at *extremely* dull parties — until Calculus II.

It's time to take polynomial division out of mothballs. In this section, I show you everything you forgot to remember about polynomial division, both with and without a remainder.

Polynomial division without a remainder

When you multiply two polynomials, you always get another polynomial. For example:

$$(x^3 + 3)(x^2 - x) = x^5 - x^4 + 3x^2 - 3x$$

Because division is the inverse of multiplication, the following equation makes intuitive sense:

$$\frac{x^5 - x^4 + 3x^2 - 3x}{x^3 + 3} = (x^2 - x)$$

Polynomial division is a reliable method for dividing one polynomial by another. It's similar to long division, so you probably won't have too much difficulty understanding it even if you've never seen it.

The best way to show you how to do polynomial division is with an example. Start with the example I've already outlined. Suppose that you want to divide $x^5 - x^4 + 3x^2 - 3x$ by $x^3 + 3$. Begin by setting up the problem as a typical long division problem (notice that I fill with zeros for the x^3 and constant terms):

$$x^3 + 3 \overline{)x^5 - x^4 + 0x^3 + 3x^2 - 3x + 0}$$

Start by focusing on the highest degree exponent in both the divisor (x^3) and dividend (x^5). Ask how many times x^3 goes into x^5 — that is, $x^5 \div x^3 = ?$ Place the answer in the quotient, and then multiply the result by the divisor as you would with long division:

$$
\begin{array}{r}
x^2 \\
x^3 + 3 \overline{)x^5 - x^4 + 0x^3 + 3x^2 - 3x + 0} \\
-(x^5 + 3x^2)
\end{array}
$$

As you can see, I multiply x^2 by x^3 to get the result of $x^5 + 3x^2$, aligning this result to keep terms of the same degree in similar columns. Next, subtract and bring down the next term, just as you would with long division:

$$
\begin{array}{r}
x^2 \\
x^3 + 3 \overline{\smash{)}\,x^5 - x^4 + 0x^3 + 3x^2 - 3x + 0} \\
\underline{-\left(x^5 + 3x^2\right)} \\
-x^4 - 3x
\end{array}
$$

Now, the cycle is complete, and you ask how many times x^3 goes into $-x^4$ — that is, $-x^4 \div x^3 = ?$ Place the answer in the quotient, and multiply the result by the divisor:

$$
\begin{array}{r}
x^2 - x \\
x^3 + 3 \overline{\smash{)}\,x^5 - x^4 + 0x^3 + 3x^2 - 3x + 0} \\
\underline{-\left(x^5 + 3x^2\right)} \\
-x^4 - 3x \\
\underline{-\left(-x^4 - 3x\right)}
\end{array}
$$

In this case, the subtraction that results works out evenly. Even if you bring down the final zero, you have nothing left to divide, which shows the following equality:

$$
\frac{x^5 - x^4 + 3x^2 - 3x}{x^3 + 3} = x^2 - x
$$

Polynomial division with a remainder

Because polynomial division looks so much like long division, it makes sense that polynomial division should, at times, leave a remainder. For example, suppose that you want to divide $x^4 - 2x^3 + 5x$ by $2x^2 - 6$:

$$
2x^2 - 6 \overline{\smash{)}\,x^4 - 2x^3 + 0x^2 + 5x + 0}
$$

This time, I fill in two zero coefficients as needed. To begin, divide x^4 by $2x^2$, multiply through, and subtract:

$$
\begin{array}{r}
\tfrac{1}{2}x^2 \\
2x^2 - 6 \overline{\smash{)}\,x^4 - 2x^3 + 0x^2 + 5x + 0} \\
\underline{-\left(x^4 - 3x^2\right)} \\
-2x^3 + 3x^2
\end{array}
$$

Don't let the fractional coefficient deter you. Sometimes polynomial division results in fractional coefficients.

Now bring down the next term ($5x$) to begin another cycle. Then, divide $-2x^3$ by $2x^2$, multiply through, and subtract:

$$2x^2 - 6 \overline{\smash{\big)}\, \begin{array}{r} \frac{1}{2}x^2 - x \\ \hline x^4 - 2x^3 + 0x^2 + 5x + 0 \\ -\left(x^4 - 3x^2\right) \\ \hline -2x^3 + 3x^2 + 5x \\ -\left(-2x^3 + 6x\right) \\ \hline 3x^2 - x \end{array}}$$

Again, bring down the next term (0) and begin another cycle by dividing $3x^2$ by $2x^2$:

$$2x^2 - 6 \overline{\smash{\big)}\, \begin{array}{r} \frac{1}{2}x^2 - x + \frac{3}{2} \\ \hline x^4 - 2x^3 + 0x^2 + 5x + 0 \\ -\left(x^4 - 3x^2\right) \\ \hline -2x^3 + 3x^2 + 5x \\ -\left(-2x^3 + 6x\right) \\ \hline 3x^2 - x + 0 \\ -\left(3x^2 - 9\right) \\ \hline -x + 9 \end{array}}$$

As with long division, the remainder indicates a fractional amount left over: the remainder divided by the divisor. So, when you have a remainder in polynomial division, you write the answer by using the following formula:

$$\text{Polynomial} = \text{Quotient} + \frac{\text{Remainder}}{\text{Divisor}}$$

If you get confused deciding how to write out the answer, think of it as a mixed number. For example, $7 \div 3 = 2$ with a remainder of 1, which you write as $2\frac{1}{3}$.

So, the polynomial division in this case provides the following equality:

$$\frac{x^4 - 2x^3 + 5x}{2x^2 - 6} = \frac{1}{2}x^2 - x + \frac{3}{2} + \frac{-x+9}{2x^2 - 6}$$

Although this result may look more complicated than the fraction you started with, you have made progress: You turned an improper rational expression (where the degree of the numerator is greater than the degree of the denominator) into a sum that includes a proper rational expression. This is similar to the practice in arithmetic of turning an improper fraction into a mixed number.

Trying out an example

In this section, I walk you through an example that takes you through just about everything in this chapter. Suppose that you want to integrate the following rational function:

$$\frac{x^4 - x^3 - 5x + 4}{(x-2)(x^2+3)}\, dx$$

This looks like a good candidate for partial fractions, as I show you earlier in this chapter in "Case #2: Distinct quadratic factors." But before you can express it as partial fractions, you need to determine whether it's proper or improper. The degree of the numerator is 4 and (because the denominator is the product of a linear and a quadratic) the degree of the entire denominator is 3. Thus, this is an improper polynomial fraction (see "Distinguishing proper and improper rational expressions" earlier in this chapter), so you can't integrate by parts.

However, you can use polynomial division to turn this improper polynomial fraction into an expression that includes a proper polynomial fraction (I omit these steps here, but I show you how earlier in this chapter in "Recalling polynomial division."):

$$\frac{x^4 - x^3 - 5x + 4}{(x-2)(x^2+3)} = x + 1 + \frac{-x^2 - 2x + 10}{(x-2)(x^2+3)}$$

As you can see, the first two terms of this expression are simple to integrate (don't forget about them!). To set up the remaining term for integration, use partial fractions:

$$\frac{-x^2 - 2x + 10}{(x-2)(x^2+3)} = \frac{A}{x-2} + \frac{Bx + C}{x^2+3}$$

Get a common denominator on the right side of the equation:

$$\frac{-x^2 - 2x + 10}{(x-2)(x^2+3)} = \frac{A(x^2+3) + (Bx + C)(x-2)}{(x-2)(x^2+3)}$$

Now multiply both sides of the equation by this denominator:

$$-x^2 - 2x + 10 = A(x^2 + 3) + (Bx + C)(x - 2)$$

Notice that $(x - 2)$ is a linear factor, so you can use the root of this factor to find the value of A. To find this value, let $x = 2$ and solve for A:

$$-(2^2) - 2(2) + 10 = A(2^2 + 3) + (B2 + C)(2 - 2)$$

$$2 - 7A$$

$$A = \frac{2}{7}$$

Substitute this value into the equation:

$$-x^2 - 2x + 10 = \frac{2}{7}(x^2 + 3) + (Bx + C)(x - 2)$$

At this point, to find the values of B and C, you need to split the equation into a system of two equations (as I show you earlier in "Working systematically with a system of equations"):

$$-x^2 - 2x + 10 = \frac{2}{7}x^2 + \frac{6}{7} + Bx^2 + Cx - 2Bx - 2C$$

$$\left(\frac{2}{7} + B + 1\right)x^2 + (-2B + C + 2)x + \left(\frac{6}{7} - 2C - 10\right) = 0$$

This splits into three equations:

$$\frac{2}{7} + B + 1 = 0$$

$$-2B + C + 2 = 0$$

$$\frac{6}{7} - 2C - 10 = 0$$

The first and the third equations show you that $B = -\frac{9}{7}$ and $C = -\frac{32}{7}$. Now you can plug the values of A, B, and C back into the sum of partial fractions:

$$\frac{2}{7(x-2)} + \frac{-9x - 32}{7(x^2 + 3)}$$

Make sure that you remember to add in the two terms $(x + 1)$ that you left behind just after you finished your polynomial division:

$$\int \frac{x^4 - x^3 - 5x + 4}{(x - 2)(x^2 + 3)}\,dx = \int \left[x + 1 + \frac{2}{7(x - 2)} + \frac{9x - 32}{7(x^2 + 3)}\right] dx$$

Thus, you can rewrite the original integral as the sum of five separate integrals:

$$\int x\,dx + \int dx + \frac{2}{7}\int \frac{1}{x - 2}\,dx - \frac{9}{7}\int \frac{x}{x^2 + 3}\,dx - \frac{32}{7}\int \frac{1}{x^2 + 3}\,dx$$

You can solve the first two of these integrals by looking at them, and the next two by variable substitution (see Chapter 5). The last is done by using the following rule:

$$\int \frac{1}{x^2+n^2}\,dx = \frac{1}{n}\arctan\frac{x}{n} + C$$

Here's the solution so that you can work the last steps yourself:

$$\frac{1}{2}x^2+x+\frac{2}{7}\ln|x-2|-\frac{9}{14}\ln|x^2+3|-\frac{32}{7\sqrt{3}}\arctan\frac{x}{\sqrt{3}} + C$$

Part III
Intermediate
Integration Topics

The 5th Wave   By Rich Tennant

"Okay, maam, I'm going to ask you to walk a straight line, then I'm going to ask you to bisect that line with a perpendicular line that slopes to the equation $y = 3x + 5$."

In this part . . .

With the basics of calculating integrals under your belt, the focus becomes using integration as a problem-solving tool. You discover how to solve more complex area problems and how to find the surface area and volume of solids.

Chapter 9

Forging into New Areas: Solving Area Problems

*W*ith your toolbox now packed with the *hows* of calculating integrals, this chapter (and Chapter 10) introduces you to some of the *whys* of calculating them.

I start with a simple rule for expressing an area as two separate definite integrals. Then I focus on improper integrals, which are integrals that are either horizontally or vertically infinite. Next, I give you a variety of practical strategies for measuring areas that are bounded by more than one function. I look at measuring areas between functions, and I also get you clear on the distinction between signed area and unsigned area.

After that, I introduce you to the Mean Value Theorem for Integrals, which provides the theoretical basis for calculating average value. Finally, I show you a formula for calculating arc length, which is the exact length between two points along a function.

Breaking Us in Two

Here's a simple but handy rule that looks complicated but is really very easy:

$$\int_a^b f(x)\,dx = \int_a^n f(x)\,dx + \int_n^b f(x)\,dx$$

This rule just says that you can split an area into two pieces, and then add up the pieces to get the area that you started with.

For example, the entire shaded area in Figure 9-1 is represented by the following integral, which you can evaluate easily:

$$\int_0^\pi \sin x\,dx$$

$$= -\cos x \Big|_{x=0}^{x=\pi}$$

$$= -\cos \pi - -\cos 0$$

$$= 1 + 1 = 2$$

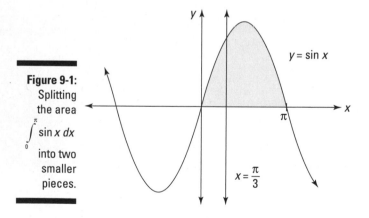

Figure 9-1:
Splitting
the area
$\int_0^\pi \sin x\,dx$
into two
smaller
pieces.

Drawing a vertical line at $x = \frac{\pi}{3}$ and splitting this area into two separate regions results in two separate integrals:

$$\int_0^{\frac{\pi}{3}} \sin x\,dx + \int_{\frac{\pi}{3}}^\pi \sin x\,dx$$

It should come as no great shock that the sum of these two smaller regions equals the entire area:

$$= -\cos x \Big|_{x=0}^{x=\frac{\pi}{3}} + -\cos x \Big|_{x=\frac{\pi}{3}}^{x=\pi}$$

$$= -\cos \frac{\pi}{3} - -\cos 0 + -\cos \pi - -\cos \frac{\pi}{3}$$

$$= \cos 0 - \cos \pi$$

$$= 1 + 1 = 2$$

Although this idea is ridiculously simple, splitting an integral into two or more integrals becomes a powerful tool for solving a variety of the area problems in this chapter.

Improper Integrals

Improper integrals come in two varieties — horizontally infinite and vertically infinite:

✔ A *horizontally infinite* improper integral contains either ∞ or –∞ (or both) as a limit of integration. See the next section, "Getting horizontal," for examples of this type of integral.

✔ A *vertically infinite* improper integral contains at least one vertical asymptote. I discuss this further in the later section "Going vertical."

Improper integrals become useful for solving a variety of problems in Chapter 10. They're also useful for getting a handle on infinite series in Chapter 12. Evaluating an improper integral is a three-step process:

1. **Express the improper integral as the limit of an integral.**

2. **Evaluate the integral by whatever method works.**

3. **Evaluate the limit.**

In this section, I show you, step by step, how to evaluate both types of improper integrals.

Getting horizontal

The first type of improper integral occurs when a definite integral has a limit of integration that's either ∞ or –∞. This type of improper integral is easy to spot because infinity is right there in the integral itself. You can't miss it.

For example, suppose that you want to evaluate the following improper integral:

$$\int_1^\infty \frac{1}{x^3}\, dx$$

Here's how you do it, step by step:

1. **Express the improper integral as the limit of an integral.**

    When the upper limit of integration is ∞, use this equation:

    $$\int_a^\infty f(x)\, dx = \lim_{c \to \infty} \int_a^c f(x)\, dx$$

    So here's what you do:

    $$\int_1^\infty \frac{1}{x^3}\, dx = \lim_{c \to \infty} \int_1^c \frac{1}{x^3}\, dx$$

2. **Evaluate the integral:**

    $$\lim_{c \to \infty} \left(-\frac{1}{2x^2}\Big|_{x=1}^{x=c} \right)$$

    $$= \lim_{c \to \infty} \left(-\frac{1}{2c^2} + \frac{1}{2} \right)$$

3. **Evaluate the limit:**

    $$= \frac{1}{2}$$

Before moving on, reflect for one moment that the area under an *infinitely* long curve is actually *finite*. Ah, the magic and power of calculus!

Similarly, suppose that you want to evaluate the following:

$$\int_{-\infty}^0 e^{5x}\, dx$$

Here's how you do it:

1. **Express the integral as the limit of an integral.**

    When the lower limit of integration is –∞, use this equation:

    $$\int_{-\infty}^b f(x)\, dx = \lim_{c \to -\infty} \int_c^b f(x)\, dx$$

    So here's what you write:

    $$\int_{-\infty}^0 e^{5x}\, dx = \lim_{c \to -\infty} \int_c^0 e^{5x}\, dx$$

2. **Evaluate the integral:**

$$= \lim_{c \to -\infty} \left(\frac{1}{5} e^{5x} \Big|_{x=c}^{x=0} \right)$$

$$= \lim_{c \to -\infty} \left(\frac{1}{5} e^0 - \frac{1}{5} e^{5c} \right)$$

$$= \lim_{c \to -\infty} \left(\frac{1}{5} - \frac{1}{5} e^{5c} \right)$$

3. **Evaluate the limit — in this case, as c approaches $-\infty$, the first term is unaffected and the second term approaches 0:**

$$= \frac{1}{5}$$

Again, calculus tells you that, in this case, the area under an infinitely long curve is finite.

Of course, sometimes the area under an infinitely long curve is infinite. In these cases, the improper integral cannot be evaluated because the limit does not exist (DNE). Here's a quick example that illustrates this situation:

$$\int_1^\infty \frac{1}{x} \, dx$$

It may not be obvious that this improper integral represents an infinitely large area. After all, the value of the function approaches 0 as x increases. But watch how this evaluation plays out:

1. **Express the improper integral as the limit of an integral:**

$$\int_1^\infty \frac{1}{x} \, dx \ell = \lim_{c \to \infty} \int_1^c \frac{1}{x} \, dx$$

2. **Evaluate the integral:**

$$= \lim_{c \to \infty} \ln x \Big|_{x=0}^{x=c}$$

$$= \lim_{c \to \infty} \ln c - \ln 1$$

At this point, you can see that the limit explodes to infinity, so it doesn't exist. Therefore, the improper integral can't be evaluated, because the area that it represents is infinite.

Going vertical

Vertically infinite improper integrals are harder to recognize than those that are horizontally infinite. An integral of this type contains at least one vertical asymptote in the area that you're measuring. (A *vertical asymptote* is a value of x where $f(x)$ equals either ∞ or $-\infty$. See Chapter 2 for more on asymptotes.) The asymptote may be a limit of integration or it may fall someplace between the two limits of integration.

Don't try to slide by and evaluate improper integrals as proper integrals. In most cases, you'll get the wrong answer!

In this section, I show you how to handle both cases of vertically infinite improper integrals.

Handling asymptotic limits of integration

Suppose that you want to evaluate the following integral:

$$\int_0^1 \frac{1}{\sqrt{x}}\, dx$$

At first glance, you may be tempted to evaluate this as a proper integral. But this function has an asymptote at $x = 0$. The presence of an asymptote at one of the limits of integration forces you to evaluate this one as an improper integral:

1. **Express the integral as the limit of an integral:**

$$\int_0^1 \frac{1}{\sqrt{x}}\, dx = \lim_{c \to 0^+} \int_c^1 \frac{1}{\sqrt{x}}\, dx$$

   Notice that in this limit, c approaches 0 from the right — that is, from the positive side — because this is the direction of approach from inside the limits of integration. (That's what the little plus sign (') in the limit in Step 2 means.)

2. **Evaluate the integral:**

   This integral is easily evaluated as $x^{-\frac{1}{2}}$, using the Power Rule as I show you in Chapter 4, so I spare you the details here:

$$= \lim_{c \to 0^+} 2\sqrt{x}\, \Big|_{x=c}^{x=1}$$

3. **Evaluate the limit:**

$$= \lim_{c \to 0^+} 2\sqrt{1} - 2\sqrt{c}$$

   At this point, direct substitution provides you with your final answer:

$$= 2$$

Piecing together discontinuous integrands

In Chapter 3, I discuss the link between integrability and continuity: If a function is continuous on an interval, it's also integrable on that interval. (Flip to Chapter 3 for a refresher on this concept.)

Some integrals that are vertically infinite have asymptotes not at the edges but someplace in the middle. The result is a *discontinuous integrand* — that is, function with a discontinuity on the interval that you're trying to integrate.

WARNING!

Discontinuous integrands are the trickiest improper integrals to spot — you really need to know how the graph of the function that you're integrating behaves. (See Chapter 2 to see graphs of the elementary functions.)

To evaluate an improper integral of this type, separate it at each asymptote into two or more integrals, as I demonstrate earlier in this chapter in "Breaking Us in Two." Then evaluate each of the resulting integrals as an improper integral, as I show you in the previous section.

For example, suppose that you want to evaluate the following integral:

$$\int_0^\pi \sec^2 x \, dx$$

Because the graph of sec x contains an asymptote at $x = \frac{\pi}{2}$ (see Chapter 2 for a view of this graph), the graph of $\sec^2 x$ has an asymptote in the same place, as you see in Figure 9-2.

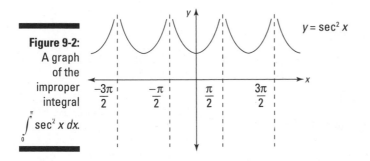

Figure 9-2: A graph of the improper integral $\int_0^\pi \sec^2 x \, dx.$

$y = \sec^2 x$

$-\frac{3\pi}{2}$ $-\frac{\pi}{2}$ $\frac{\pi}{2}$ $\frac{3\pi}{2}$

To evaluate this integral, break it into two integrals at the value of x where the asymptote is located:

$$\int_0^\pi \sec^2 x \, dx = \int_0^{\frac{\pi}{2}} \sec^2 x \, dx + \int_{\frac{\pi}{2}}^\pi \sec^2 x \, dx$$

Now, evaluate the sum of the two resulting integrals.

TIP

You can save yourself a lot of work by noticing when two regions are symmetrical. In this case, the asymptote at $x = \frac{\pi}{2}$ splits the shaded area into two symmetrical regions. So you can find one integral and then double it to get your answer:

$$= 2 \int_{0}^{\frac{\pi}{2}} \sec^2 x \, dx$$

Now, use the steps from the previous section to evaluate this integral:

1. **Express the integral as the limit of an integral:**

$$= 2 \lim_{c \to \frac{\pi}{2}} \int_{0}^{c} \sec^2 x \, dx$$

In this case, the vertical asymptote is at the upper limit of integration, so c approaches $\frac{\pi}{2}$ from the left — that is, from inside the interval where you're measuring the area.

2. **Evaluate the integral:**

$$= 2 \lim_{c \to \frac{\pi}{2}} \left(\tan x \big|_{x=0}^{x=c} \right)$$

$$= 2 \lim_{c \to \frac{\pi}{2}} \left(\tan c - \tan 0 \right)$$

3. **Evaluate the limit:**

Note that $\tan \frac{\pi}{2}$ is undefined, because the function $\tan x$ has an asymptote at $x = \frac{\pi}{2}$, so the limit does not exist (DNE). Therefore, the integral that you're trying to evaluate also does not exist because the area that it represents is infinite.

Solving Area Problems with More Than One Function

The definite integral allows you to find the signed area under any interval of a single function. But when you want to find an area defined by more than one function, you need to be creative and piece together a solution. Professors love these problems as exam questions, because they test your reasoning skills as well as your calculus knowledge.

Fortunately, when you approach problems of this type correctly, you find that they're not terribly difficult. The trick is to break down the problem into two or more regions that you can measure by using the definite integral, and then use addition or subtraction to find the area that you're looking for.

In this section, I get you up to speed on problems that involve more than one definite integral.

Finding the area under more than one function

Sometimes, a single geometric area is described by more than one function. For example, suppose that you want to find the shaded area shown in Figure 9-3.

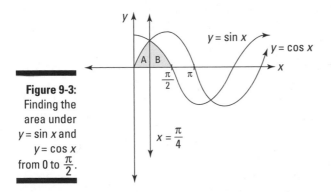

Figure 9-3: Finding the area under $y = \sin x$ and $y = \cos x$ from 0 to $\frac{\pi}{2}$.

The first thing to notice is that the shaded area isn't under a single function, so you can't expect to use a single integral to find it. Instead, the region labeled A is under $y = \sin x$ and the region labeled B is under $y = \cos x$. First, set up an integral to find the area of both of these regions:

$$A = \int_{0}^{\frac{\pi}{4}} \sin x \, dx$$

$$B = \int_{\frac{\pi}{4}}^{\frac{\pi}{2}} \cos x \, dx$$

Now, set up an equation to find their combined area:

$$A + B = \int_{0}^{\frac{\pi}{4}} \sin x \, dx + \int_{\frac{\pi}{4}}^{\frac{\pi}{2}} \cos x \, dx$$

At this point, you can evaluate each of these integrals separately. But there's an easier way.

Because region A and region B are symmetrical, they have the same area. So you can find their combined area by doubling the area of a single region:

$$= 2A = 2 \int_{0}^{\frac{\pi}{4}} \sin x \, dx$$

I choose to double region A because the integral limits of integration are easier, but doubling region B also works. Now, integrate to find your answer:

$$= 2 \left(-\cos x \right) \Big|_{x=0}^{x=\frac{\pi}{4}}$$

$$= 2 \left(-\cos \frac{\pi}{4} - -\cos 0 \right)$$

$$= 2 \left(-\frac{\sqrt{2}}{2} + 1 \right)$$

$$= 2 - \sqrt{2} \approx 0.586$$

Finding the area between two functions

To find an area between two functions, you need to set up an equation with a combination of definite integrals of both functions. For example, suppose that you want to calculate the shaded area in Figure 9-4.

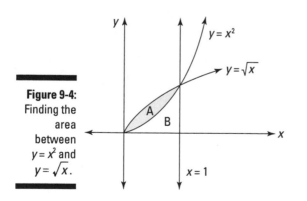

Figure 9-4:
Finding the area between $y = x^2$ and $y = \sqrt{x}$.

First, notice that the two functions $y = x^2$ and $y = \sqrt{x}$ intersect where $x = 1$. This is important information because it enables you to set up two definite integrals to help you find region A:

$$A + B = \int_0^1 \sqrt{x}\, dx$$

$$B = \int_0^1 x^2\, dx$$

Although neither equation gives you the exact information that you're look-ing for, together they help you out. Just subtract the second equation from the first as follows:

$$A = A + B - B = \int_0^1 \sqrt{x}\, dx - \int_0^1 x^2\, dx$$

With the problem set up properly, now all you have to do is evaluate the two integrals:

$$= \left(\frac{2}{3} x^{\frac{3}{2}} \Big|_{x=0}^{x=1} \right) - \left(-\frac{1}{3} x^3 \Big|_{x=0}^{x=1} \right)$$

$$= \left(\frac{2}{3} - 0 \right) - \left(\frac{1}{3} - 0 \right) = \frac{1}{3}$$

So the area between the two curves is $\frac{1}{3}$.

As another example, suppose that you want to find the shaded area in Figure 9-5.

This time, the shaded area is two separate regions, labeled A and B. Region A is bounded above by $y = x^{\frac{1}{3}}$ and bounded below by $y = x$. However, for region B, the situation is reversed, and the region is bounded above by $y = x$ and bounded below by $y = x^{\frac{1}{3}}$. I also label region C and region D, both of which figure into the problem.

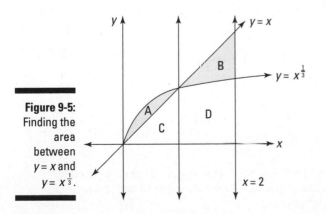

Figure 9-5:
Finding the
area
between
$y = x$ and
$y = x^{\frac{1}{3}}$.

The first important step is finding where the two functions intersect — that is, where the following equation is true:

$$x = x^{\frac{1}{3}}$$

Fortunately, it's easy to see that $x = 1$ satisfies this equation.

Now, you want to build a few definite integrals to help you find the areas of region A and region B. Here are two that can help with region A:

$$A + C = \int_0^1 x^{\frac{1}{3}}\, dx$$

$$C = \int_0^1 x\, dx = \frac{1}{2}$$

Notice that I evaluate the second definite integral *without* calculus, using simple geometry as I show you in Chapter 1. This is perfectly valid and a great timesaver.

Subtracting the second equation from the first provides an equation for the area of region A:

$$A = A + C - C = \int_0^1 x^{\frac{1}{3}}\, dx - \frac{1}{2}$$

Now, build two definite integrals to help you find the area of region B:

$$B + D = \int_1^2 x\, dx = \frac{3}{2}$$

$$D = \int_1^2 x^{\frac{1}{3}}\, dx$$

This time, I evaluate the first definite integral by using geometry instead of calculus. Subtracting the second equation from the first gives an equation for the area of region B:

$$B = B + D - D = \frac{3}{2} - \int_1^2 x^{\frac{1}{3}}\, dx$$

Now you can set up an equation to solve the problem:

$$A + B = \int_0^1 x^{\frac{1}{3}}\,dx - \frac{1}{2} + \frac{3}{2} - \int_1^2 x^{\frac{1}{3}}\,dx$$

$$= \int_0^1 x^{\frac{1}{3}}\,dx - \int_1^2 x^{\frac{1}{3}}\,dx + 1$$

At this point, you're forced to do some calculus:

$$= \left(\frac{3}{4} x^{\frac{4}{3}} \Big|_{x=0}^{x=1} \right) - \left(\frac{3}{4} x^{\frac{4}{3}} \Big|_1^2 + 1 \right)$$

$$= \left(\frac{3}{4}(1)^{\frac{4}{3}} - 0 \right) - \left(\frac{3}{4}(2)^{\frac{4}{3}} - \frac{3}{4}(1)^{\frac{4}{3}} \right) + 1$$

The rest is just arithmetic:

$$= \frac{3}{4} - \frac{3}{4}(16)^{\frac{1}{3}} + \frac{3}{4} + 1$$

$$= \frac{5}{2} - \frac{3}{4}(16)^{\frac{1}{3}}$$

$$\approx 0.6101$$

Looking for a sign

The solution to a definite integral gives you the *signed* area of a region (see Chapter 3 for more). In some cases, signed area is what you want, but in some problems you're looking for *unsigned* area.

The signed area above the *x*-axis is positive, but signed area below the *x*-axis is negative. In contrast, unsigned area is *always* positive. The concept of unsigned area is similar to the concept of absolute value. So, if it's helpful, think of unsigned area as the absolute value of a definite integral.

In problems where you're asked to find the area of a shaded region on a graph, you're looking for unsigned area. But if you're unsure whether a question is asking you to find signed or unsigned area, ask the professor. This goes double if an exam question is unclear. Most professors will answer clarifying questions, so don't be shy to ask.

For example, suppose that you're asked to calculate the shaded unsigned area that's shown in Figure 9-6.

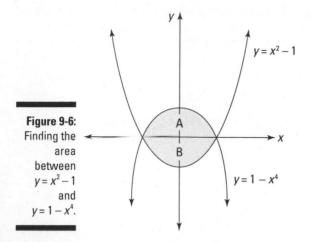

This area is actually the sum of region A, which is above the x-axis, and region B, which is below it. To solve the problem, you need to find the sum of the unsigned areas of these two regions.

Fortunately, both functions intersect each other and the x-axis at the same two values of x: $x = -1$ and $x = 1$. Set up definite integrals to find the area of each region as follows:

$$A = \int_{-1}^{1} \left(1 - x^4\right) dx$$

$$B = -\int_{-1}^{1} \left(x^2 - 1\right) dx$$

Notice that I negate the definite integral for region B to account for the fact that the definite integral produces negative area below the x-axis. Now, just add the two equations together:

$$A + B = \int_{-1}^{1} \left(1 - x^4\right) dx - \int_{-1}^{1} \left(x^2 - 1\right) dx$$

Solving this equation gives you the answer that you're looking for (be careful with all those minus signs!):

$$= \left(x - \frac{1}{5}x^5 \Big|_{x=-1}^{x=1}\right) - \left(\frac{1}{3}x^3 - x \Big|_{x=-1}^{x=1}\right)$$

$$= \left[\left(1 - \frac{1}{5}\right) - \left(-1 - -\frac{1}{5}\right)\right] - \left[\left(\frac{1}{3} - 1\right) - \left(-\frac{1}{3} - -1\right)\right]$$

$$= \frac{4}{5} + \frac{4}{5} - \left(-\frac{2}{3} - \frac{2}{3}\right)$$

Notice at this point that the expression in the parentheses — representing the signed area of region B — is negative. But the minus sign outside the parentheses automatically flips the sign as intended:

$$= \frac{8}{5} + \frac{4}{3} = \frac{44}{15}$$

Measuring unsigned area between curves with a quick trick

After you understand the concept of measuring unsigned area (which I discuss in the previous section), you're ready for a trick that makes measuring the area between curves very straightforward. As I say earlier in this chapter, professors love to stick these types of problem on exams. So here's a difficult exam question that's worth spending some time with:

> Find the unsigned shaded area in Figure 9-7. Approximate your answer to two decimal places by using cos 4 = –0.65.

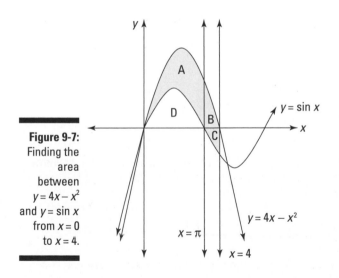

Figure 9-7: Finding the area between $y = 4x - x^2$ and $y = \sin x$ from $x = 0$ to $x = 4$.

The first step is to find an equation for the solution (which will probably give you partial credit), and then worry about solving it.

I split the shaded area into three regions labeled A, B, and C. I also label region D, which you need to consider. Notice that $x = \pi$ separates regions A and B, and the x-axis separates regions B and C.

You could find three separate equations for regions A, B, and C, but there's a better way.

TIP

To measure the unsigned area between two functions, use this quick trick:

Area = Integral of Top Function – Integral of Bottom Function

That's it! Instead of measuring the area above and below the x-axis, just plug the two integrals into this formula. In this problem, the top function is $4x - x^2$ and the bottom function is $\sin x$:

$$= \int_0^4 \left(4x - x^2\right) dx - \int_0^4 \sin x\, dx$$

This evaluation isn't too horrible:

$$\left(2x^2 - \frac{1}{3}x^3 \Big|_{x=0}^{x=4}\right) - \left(-\cos x \Big|_{x=0}^{x=4}\right)$$

$$= \left[\left(2(4)^2 - \frac{1}{3}(4)^3\right) - 0\right] - \left(-\cos 4 - -\cos 0\right)$$

$$= 32 - \frac{64}{3} + \cos 4 - 1$$

When you get to this point, you can already see that you're on track, because the professor was nice enough to give you an approximate value for cos 4:

$$\approx 32 - 21.33 - 0.65 - 1 = 9.02$$

So the unsigned area between the two functions is approximately 9.02 units.

REMEMBER

If the two functions change positions — that is, the top becomes the bottom and the bottom becomes the top — you may need to break the problem up into regions, as I show you earlier in this chapter. But even in this case, you can still save a lot of time by using this trick.

For example, earlier in this chapter, in "Finding the area between two functions," I measure the shaded area from Figure 9-5 by using four separate regions. Here's how to do it using the trick in this section.

Notice that the two functions cross at $x = 1$. So, from 0 to 1, the top function is $x^{\frac{1}{3}}$ and from 1 to 2 the top function is x. So, set up two separate equations, one for region A and another for the region B:

$$A = \int_0^1 x^{\frac{1}{3}}\, dx - \int_0^1 x\, dx$$

$$B = \int_1^2 x\, dx - \int_1^2 x^{\frac{1}{3}}\, dx$$

When the calculations are complete, you get the following values for A and B:

$$A = \frac{1}{4}$$

$$B = \frac{9}{4} - \frac{3}{4}(16)^{\frac{1}{3}}$$

Add these two values together to get your answer:

$$A + B = \frac{5}{2} - \frac{3}{4}(16)^{\frac{1}{3}} \approx 0.6101$$

As you can see, the top-and-bottom trick gets you the same answer much more simply than measuring regions.

The Mean Value Theorem for Integrals

The *Mean Value Theorem for Integrals* guarantees that for every definite integral, a rectangle with the same area and width exists. Moreover, if you superimpose this rectangle on the definite integral, the top of the rectangle intersects the function. This rectangle, by the way, is called the *mean-value rectangle* for that definite integral. Its existence allows you to calculate the *average value* of the definite integral.

Calculus boasts *two* Mean Value Theorems — one for derivatives and one for integrals. This section discusses the Mean Value Theorem for Integrals. You can find out about the Mean Value Theorem for Derivatives in *Calculus For Dummies* by Mark Ryan (Wiley).

The best way to see how this theorem works is with a visual example. The first graph in Figure 9-8 shows the region described by the definite integral $A = \int_0^1 x^{\frac{1}{3}}\, dx - \int_0^1 x\, dx$. This region obviously has a width of 1, and you can evaluate it easily to show that its area is $\frac{7}{3}$.

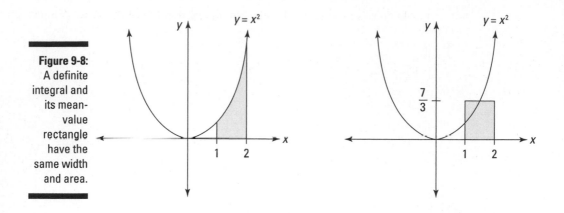

The second graph in Figure 9-8 shows a rectangle with a width of 1 and an area of $\frac{7}{3}$. It should come as no surprise that this rectangle's height is also $\frac{7}{3}$, so the top of this rectangle intersects the original function.

The fact that the top of the mean-value rectangle intersects the function is mostly a matter of common sense. After all, the height of this rectangle represents the average value that the function attains over a given interval. This value must fall someplace between the function's maximum and minimum values on that interval.

Here's the formal statement of the Mean Value Theorem for Integrals:

If $f(x)$ is a continuous function on the closed interval $[a, b]$, then there exists a number c in that interval such that:

$$\int_a^b f(x)\,dx = f(c) \cdot (b-a)$$

This equation may look complicated, but it's basically a restatement of this familiar equation for the area of a rectangle:

Area = Height · Width

In other words: Start with a definite integral that expresses an area, and then draw a rectangle of equal area with the same width $(b - a)$. The height of that rectangle — $f(c)$ — is such that its top edge intersects the function where $x = c$.

The value $f(c)$ is the *average value* of $f(x)$ over the interval $[a, b]$. You can calculate it by rearranging the equation stated in the theorem:

$$f(c) = \frac{1}{b-a} \cdot \int_a^b f(x)\,dx$$

For example, here's how you calculate the average value of the shaded area in Figure 9-9:

$$f(c) = \frac{1}{4-2} \cdot \int_{2}^{4} x^3 \, dx$$

$$= \frac{1}{2} \left(\frac{1}{4} x^4 \Big|_{x=2}^{x=4} \right)$$

$$= \frac{1}{2} \left(\frac{1}{4} 4^4 - \frac{1}{4} 2^4 \right)$$

$$= \frac{1}{2} (64 - 4) = 30$$

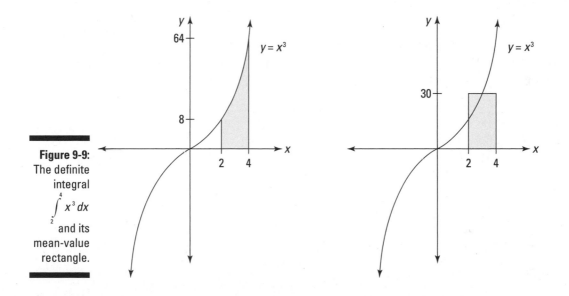

Figure 9-9:
The definite integral
$$\int_{2}^{4} x^3 \, dx$$
and its mean-value rectangle.

Not surprisingly, the average value of this integral is 30, a value between the function's minimum of 8 and its maximum of 64.

Calculating Arc Length

The arc length of a function on a given interval is the length from the starting point to the ending point as measured along the graph of that function.

In a sense, arc length is similar to the practical measurement of driving distance. For example, you may live only 5 miles from work "as the crow flies," but when you check your odometer, you may find that the actual drive is closer to 7 miles. Similarly, the straight-line distance between two points is always less than the arc length along a curved function that connects them.

Using the formula, however, often involves trig substitution (see Chapter 7 for a refresher on this method of integration).

The formula for the arc length along a function $y = f(x)$ from a to b is as follows:

$$\int_a^b \sqrt{1 + \left(\frac{dy}{dx}\right)^2}\, dx$$

For example, suppose that you want to calculate the arc length along the function $y = x^2$ from the point where $x = 0$ to the point where $x = 2$ (see Figure 9-10).

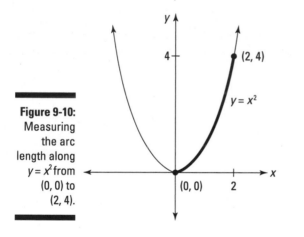

Figure 9-10:
Measuring the arc length along $y = x^2$ from (0, 0) to (2, 4).

Before you begin, notice that if you draw a straight line between these two points, (0, 0) and (2, 4), its length is $\sqrt{20} \approx 4.4721$. So the arc length should be slightly greater.

To calculate the arc length, first find the derivative of the function x^2:

$$\frac{dy}{dx} = 2x$$

Now, plug this derivative and the limits of integration into the formula as follows:

$$\int_0^2 \sqrt{1 + (2x)^2}\, dx$$

$$= \int_0^2 \sqrt{1 + 4x^2}\, dx$$

Calculating arc length usually gives you an opportunity to practice trig substitution — in particular, the tangent case. When you draw your trig substitution triangle, place $\sqrt{1 + 4x^2}$ on the hypotenuse, $2x$ on the opposite side, and 1 on the adjacent side. This gives you the following substitutions:

$$\sqrt{1 + 4x^2} = \sec\theta$$

$$2x = \tan\theta$$

$$x = \frac{1}{2}\tan\theta$$

$$dx = \frac{1}{2}\sec^2\theta\, d\theta$$

The result is this integral:

$$\frac{1}{2}\int \sec^3\theta\, d\theta$$

Notice that I remove the limits of integration because I plan to change the variable back to x before computing the definite integral. I spare you the details of calculating this indefinite integral, but you can see them in Chapter 7. Here's the result:

$$= \frac{1}{4}\left(\ln\,|\sec\theta + \tan\theta| + \tan\theta\sec\theta\right) + C$$

Now, write the each $\sec\theta$ and $\tan\theta$ in terms of x:

$$\frac{1}{4}\left(\ln\left|\sqrt{1 + 4x^2} + x^2 + 2x\right| + 2x\sqrt{1 + 4x^2}\right) + C$$

At this point, I'm ready to evaluate the definite integral that I leave off earlier:

$$\int_0^2 \sqrt{1 + 4x^2}\, dx$$

$$= \frac{1}{4}\left(\ln\left|\sqrt{1 + 4x^2} + 2x\right| + 2x\sqrt{1 + 4x^2}\right)\Big|_{x=0}^{x=2}$$

$$= \frac{1}{4}\left(\ln\left|\sqrt{1 + 4\,(2)^2} + 2\,(2)\right| + 2\,(2)\sqrt{1 + 4\,(2)^2}\right) - 0$$

You can either take my word that the second part of this substitution works out to 0 or calculate it yourself. To finish up:

$$= \frac{1}{4}\ln\left|\sqrt{17} + 4\right| + \sqrt{17}$$

$$\approx 0.5236 + 4.1231 = 4.6467$$

Chapter 10

Pump up the Volume: Using Calculus to Solve 3-D Problems

. . . .

In This Chapter

▶ Understanding the meat-slicer method for finding volume

▶ Using inverses to make a problem easier to solve

▶ Solving problems with solids of revolution and surfaces of revolution

▶ Finding the space between two surfaces

▶ Understanding the shell method for finding volume

. . . .

*I*n Chapter 9, I show you a bunch of different ways to use integrals to find area. In this chapter, you add a dimension by discovering how to use integrals to find volumes and surface areas of solids.

First, I show you how to find the volume of a solid by using the meat-slicer method, which is really a 3-D extension of the basic integration tactic you already know from Chapter 1: slicing an area into an infinite number of pieces and adding them up.

As with a real meat slicer, this method works best when the blade is slicing vertically — that is, perpendicular to the *x*-axis. So, I also show you how to use inverses to rotate some solids into the proper position.

After that, I show you how to solve two common types of problems that calculus teachers just love: finding the volume of a solid of revolution and finding the area of a surface of revolution.

With these techniques in your back pocket, you move on to more complex problems, where a solid is described as the space between two surfaces. These problems are the 3-D equivalent of finding an area between two curves, which I discuss in Chapter 9.

To finish up, I give you an additional way to find the volume of a solid: the shell method. Then, I provide some practical perspective on all the methods in the chapter so you know when to use them.

Slicing Your Way to Success

Did you ever marvel at the way in which a meat slicer turns an entire salami into dozens of tasty little paper-thin circles? Even if you're a vegetarian, calculus provides you with an animal-friendly alternative: the meat-slicer method for measuring the volume of solids.

The *meat-slicer method* works best with solids that have similar cross sections. (I discuss this further in the following section.) Here's the plan:

1. **Find an expression that represents the area of a random cross section of the solid in terms of *x*.**

2. **Use this expression to build a definite integral (in terms of *dx*) that represents the volume of the solid.**

3. **Evaluate this integral.**

Don't worry if these steps don't make a whole lot of sense yet. In this section, I show you when and how to use the meat-slicer method to find volumes that would be difficult or impossible without calculus.

Finding the volume of a solid with congruent cross sections

Before I get into calculus, I want to provide a little bit of background on finding the volume of solids. Spending a few minutes thinking about how volume is measured *without* calculus pays off big-time when you step into the calculus arena. This is strictly no-brainer stuff — some basic, solid geometry that you probably know already. So just lie back and coast through this section.

One of the simplest solids to find the volume of is a prism. A *prism* is a solid that has all congruent cross sections in the shape of a polygon. That is, no matter how you slice a prism parallel to its base, its cross section is the same shape and area as the base itself.

The formula for the volume of a prism is simply the area of the base times the height:

$$V = A_b \cdot h$$

So, if you have a triangular prism with a height of 3 inches and a base area of 2 square inches, its volume is 6 cubic inches.

This formula also works for cylinders — which are sort of prisms with a circular base — and generally any solid that has congruent cross sections. For example, the odd-looking solid in Figure 10-1 fits the bill nicely. In this case, you're given the information that the area of the base is 7 cm^2 and the height is 4 cm, so the volume of this solid is 28 cm^3.

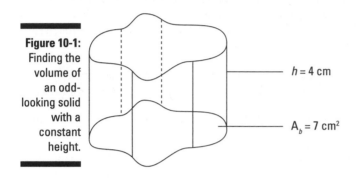

Figure 10-1:
Finding the volume of an odd-looking solid with a constant height.

$h = 4$ cm

$A_b = 7$ cm^2

Finding the volume of a solid with congruent cross sections is always simple as long as you know two things:

✔ The area of the base — that is, the area of any cross section

✔ The height of the solid

Finding the volume of a solid with similar cross sections

In the previous section, you didn't have to use any calculus brain cells. But now, suppose that you want to find the volume of the scary-looking hyperbolic cooling tower on the left side of Figure 10-2.

What makes this problem out of the reach of the formula for prisms and cylinders? In this case, slicing parallel to the base always results in the same shape — a circle — but the area may differ. That is, the solid has *similar* cross sections rather than congruent ones.

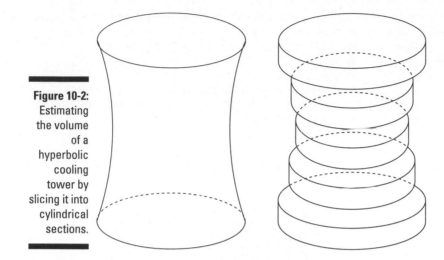

Figure 10-2:
Estimating
the volume
of a
hyperbolic
cooling
tower by
slicing it into
cylindrical
sections.

You can estimate this volume by slicing the solid into numerous cylinders, finding the volume of each cylinder by using the formula for constant-height solids, and adding these separate volumes. Of course, making more slices improves your estimate. And, as you may already suspect, adding the limit of an infinite number of slices gives you the exact volume of the solid.

Hmmm . . . this is beginning to sound like a job for calculus. In fact, what I hint at in this section is the meat-slicer method, which works well for measuring solids that have similar cross sections.

When a problem asks you to find the volume of a solid, look at the picture of this solid and figure out how to slice it up so that all the cross sections are similar. This is a good first step in understanding the problem so that you can solve it.

To measure weird-shaped solids that *don't* have similar cross sections, you need multivariable calculus, which is the subject of Calculus III. See Chapter 14 for an overview of this topic.

Measuring the volume of a pyramid

Suppose that you want to find the volume of a pyramid with a 6-x-6-unit square base and a height of 3 units. Geometry tells you that you can use the following formula:

$$V = \frac{1}{3}bh = \frac{1}{3}(36)(3) = 36$$

This formula works just fine, but it doesn't give you insight into how to solve similar problems; it works only for pyramids. The meat-slicer method, however, provides an approach to the problem that you can generalize to use for many other types of solids.

To start out, I skewer this pyramid on the *x*-axis of a graph, as shown in Figure 10-3. Notice that the vertex of the pyramid is at the origin, and the center of the base is at the point (6, 0).

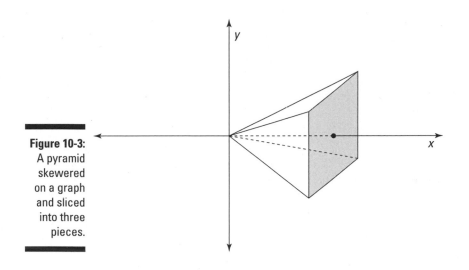

Figure 10-3:
A pyramid
skewered
on a graph
and sliced
into three
pieces.

To find the exact volume of the pyramid, here's what you do:

1. **Find an expression that represents the area of a random cross section of the pyramid in terms of *x*.**

   At *x* = 1, the cross section is 2^2 = 4. At *x* = 2, it's 4^2 = 16. And at *x* = 3, it's 6^2 = 36. So generally speaking, the area of the cross section is:

   $$A = (2x)^2 = 4x^2$$

2. **Use this expression to build a definite integral that represents the volume of the pyramid.**

   In this case, the limits of integration are 0 and 3, so:

   $$V = \int_0^3 4x^2\,dx$$

3. **Evaluate this integral:**

   $$\frac{4}{3}x^3\Big|_{x=0}^{x=3}$$
   $$= \frac{4}{3}\,3^3 - 0 = 36$$

This is the same answer provided by the formula for the pyramid. But this method can be applied to a far wider variety of solids.

Measuring the volume of a weird solid

After you know the basic meat-slicer technique, you can apply it to any solid with a cross section that's a function of x. In some cases, these solids are harder to describe than they are to measure. For example, have a look at Figure 10-4.

The solid in Figure 10-4 consists of two exponential curves — one described by the equation $y = e^x$, and the other described by placing the same curve directly in front of the x-axis — joined by straight lines. The other sides of the solid are bounded planes slicing perpendicularly in a variety of directions.

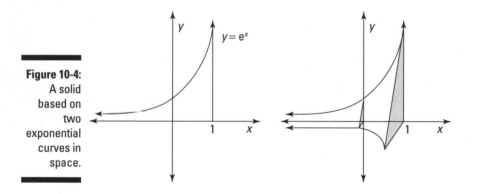

Figure 10-4:
A solid based on two exponential curves in space.

Notice that when you slice this solid perpendicular with the x-axis, its cross section is always an isosceles right triangle. This is an easy shape to measure, so the slicing method works nicely to measure the volume of this solid. Here are the steps:

1. **Find an expression that represents the area of a random cross section of the solid.**

   The triangle on the y-axis has a height and base of 1 — that is, e^0. And the triangle on the line $x = 1$ has a height and base of e^1, which is e. In general, the height and base of any cross section triangle is e^x.

   So, here's how to use the formula for the area of a triangle to find the area of a cross section in terms of x:

   $$A = \frac{1}{2}b \cdot h = \frac{1}{2}e^x \cdot e^x = \frac{1}{2}e^{2x}$$

2. Use this expression to build a definite integral that represents the volume of the solid.

Now that you know how to measure the area of a cross section, integrate to add all the cross sections from $x = 0$ to $x = 1$:

$$V = \int_0^1 \frac{1}{2} e^{2x} \, dx$$

3. Evaluate this integral to find the volume.

$$= \frac{1}{2} \int_0^1 e^{2x} \, dx$$

$$= \frac{1}{4} e^{2x} \Big|_{x=0}^{x=1}$$

$$= \frac{1}{4} e^2 - \frac{1}{4} e^0$$

$$\approx 1.597$$

Turning a Problem on Its Side

When using a real meat slicer, you need to find a way to turn whatever you're slicing on its side so that it fits. The same is true for calculus problems.

For example, suppose that you want to measure the volume of the solid shown in Figure 10-5.

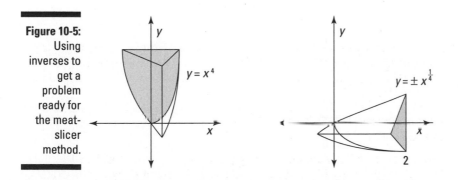

The good news is that this solid has cross sections that are all similar triangles, so the meat-slicer method will work. Unfortunately, as the problem currently stands, you'd have to make your slices perpendicular to the y-axis. But to use the meat-slicer method, you must make your slices perpendicular to the x-axis.

To solve the problem, you first need to flip the solid over to the x-axis, as shown on the right side of Figure 10-5. The easiest way to do this is to use the inverse of the function $y = x^4$. To find the inverse, switch x and y in the equation and solve for y:

$$x = y^4$$
$$\pm\left(x^{\frac{1}{4}}\right) = y$$

Note that the resulting equation $\pm\left(x^{\frac{1}{4}}\right) = y$ in this case isn't a function of x because a single x-value can produce more than one y-value. However, you can use this equation in conjunction with the meat-slicer method to find the volume that you're looking for.

1. **Find an expression that represents the area of a random cross section of the solid.**

   The cross section is an isosceles triangle with a height of 3 and a base of $2x^{\frac{1}{4}}$, so use the formula for the area of a triangle:

   $$A = \frac{1}{2}\,bh = \frac{1}{2}\left(2x^{\frac{1}{4}}\right)(3) = 3x^{\frac{1}{4}}$$

2. **Use this expression to build a definite integral that represents the volume of the solid.**

   $$V = \int_0^2 3x^{\frac{1}{4}}\,dx$$

3. **Solve the integral.**

   $$3\left(\frac{4}{5}\right)x^{\frac{5}{4}}\Big|_{x=0}^{x=2}$$

   $$\frac{12}{5}\,x^{\frac{5}{4}}\Big|_{x=0}^{x=2}$$

   Now, evaluate this expression:

   $$= \frac{12}{5}\,2^{\frac{5}{4}} - 0$$

   $$= \frac{12}{5}\,32^{\frac{1}{4}}$$

   $$\approx 5.7082$$

Two Revolutionary Problems

Calculus professors are always on the lookout for new ways to torture their students. Okay, that's a slight exaggeration. Still, sometimes it's hard to fathom exactly why a problem without much practical use makes the Calculus Hall of Fame.

In this section, I show you how to tackle two problems of dubious practical value (unless you consider the practicality passing Calculus II!). First, I show you how to find the volume of a *solid of revolution:* a solid created by spinning a function around an axis. The meat-slicer method, which I discuss in the previous section, also applies to problems of this kind.

Next, I show you how to find the area of a *surface of revolution:* a surface created by spinning a function around an axis. Fortunately, a formula exists for finding or solving this type of problem.

Solidifying your understanding of solids of revolution

A solid of revolution is created by taking a function, or part of a function, and spinning it around an axis — in most cases, either the *x*-axis or the *y*-axis.

For example, the left side of Figure 10-6 shows the function $y = 2 \sin x$ between $x = 0$ and $x = \frac{\pi}{2}$.

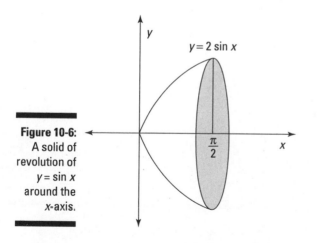

Figure 10-6:
A solid of revolution of $y = \sin x$ around the *x*-axis.

Every solid of revolution has circular cross sections perpendicular to the axis of revolution. When the axis of revolution is the *x*-axis (or any other line that's parallel with the *x*-axis), you can use the meat-slicer method directly, as I show you earlier in this chapter.

However, when the axis of revolution is the y-axis (or any other line that's parallel with the y-axis), you need to modify the problem as I show you in the earlier section "Turning a Problem on Its Side."

To find the volume of this solid of revolution, use the meat-slicer method:

1. **Find an expression that represents the area of a random cross section of the solid (in terms of x).**

   This cross section is a circle with a radius of 2 sin x:

   $$A = \pi r^2 = \pi (2 \sin x)^2 = 4\pi \sin^2 x$$

2. **Use this expression to build a definite integral (in terms of dx) that represents the volume of the solid.**

   This time, the limits of integration are from 0 to $\frac{\pi}{2}$:

   $$V = \int_0^{\frac{\pi}{2}} 4\pi \sin^2 x \, dx$$

   $$= 4\pi \int_0^{\frac{\pi}{2}} \sin^2 x \, dx$$

3. **Evaluate this integral by using the half-angle formula for sines, as I show you in Chapter 7:**

   $$= 4\pi \int_0^{\frac{\pi}{2}} \frac{(1 - \cos 2x)}{2} \, dx$$

   $$= 2\pi \left(\int_0^{\frac{\pi}{2}} 1 \, dx - \int_0^{\frac{\pi}{2}} \cos 2x \, dx \right)$$

   $$= 2\pi \left(x \Big|_{x=0}^{x=\frac{\pi}{2}} - \frac{1}{2} \sin 2x \Big|_{x=0}^{x=\frac{\pi}{2}} \right)$$

   Now, evaluate:

   $$= 2\pi \left[\left(\frac{\pi}{2} - 0 \right) - \left(\frac{1}{2} \sin \pi - 0 \right) \right]$$

   $$= 2\pi \left(\frac{\pi}{2} \right)$$

   $$= \pi^2$$

   $$\approx 9.8696$$

So the volume of this solid of revolution is approximately 9.8696 cubic units.

Later in this chapter, I give you more practice measuring the volume of solids of revolution.

Skimming the surface of revolution

The nice thing about finding the area of a surface of revolution is that there's a formula you can use. Memorize it and you're halfway done.

To find the area of a surface of revolution between a and b, use the following formula:

$$A = \int_a^b 2\pi r \sqrt{1 + \left(\frac{dy}{dx}\right)^2}\, dx$$

This formula looks long and complicated, but it makes more sense when you spend a minute thinking about it. The integral is made from two pieces:

- ✔ The arc-length formula, which measures the length along the surface (see Chapter 9)
- ✔ The formula for the circumference of a circle, which measures the length around the surface

So multiplying these two pieces together is similar to multiplying length and width to find the area of a rectangle. In effect, the formula allows you to measure surface area as an infinite number of little rectangles.

When you're measuring the surface of revolution of a function $f(x)$ around the x-axis, substitute $r = f(x)$ into the formula I gave you:

$$A = \int_a^b 2\pi f(x) \sqrt{1 + \left[f'(x)\right]^2}\, dx$$

For example, suppose that you want to find the surface of revolution that's shown in Figure 10-7.

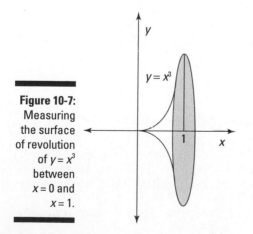

Figure 10-7: Measuring the surface of revolution of $y = x^3$ between $x = 0$ and $x = 1$.

To solve this problem, first note that for $f(x) = x^3$, $f'(x) = 3x^2$. So set up the problem as follows:

$$A = \int_0^1 2\pi x^3 \sqrt{1 + (3x^2)^2}\, dx$$

To start off, simplify the problem a bit:

$$-2\pi \int_0^1 x^3 \sqrt{1 + 9x^4}\, dx$$

You can solve this problem by using variable substitution:

Let $u = 1 + 9x^4$

$du = 36x^3\, dx$

$$= \frac{1}{36} \cdot 2\pi \int_1^{10} \sqrt{u}\, du$$

Notice that I change the limits of integration: When $x = 0$, $u = 1$. And when $x = 1$, $u = 10$.

$$= \frac{1}{18} \pi \int_1^{10} \sqrt{u}\, du$$

Now, you can perform the integration:

$$= \frac{1}{18} \pi \cdot \frac{2}{3} u^{\frac{3}{2}} \Big|_{u=1}^{u=10}$$

$$= \frac{1}{27} \pi u^{\frac{3}{2}} \Big|_{u=1}^{u=10}$$

Finally, evaluate the definite integral:

$$= \frac{1}{27} \pi\, 10^{\frac{3}{2}} - \frac{1}{27} \pi\, 1^{\frac{3}{2}}$$

$$= \frac{1}{27} \pi\, 10 \sqrt{10} - \frac{1}{27} \pi$$

$$\approx 3.5631$$

Finding the Space Between

In Chapter 9, I show you how to find the area between two curves by subtracting one integral from another. This same principle applies in three dimensions to find the volume of a solid that falls between two different surfaces.

The meat-slicer method, which I describe earlier in this chapter, is useful for many problems of this kind. The trick is to find a way to describe the donut-shaped area of a cross section as the difference between two integrals: one integral that describes the whole shape minus another that describes the hole.

For example, suppose that you want to find the volume of the solid shown in Figure 10-8.

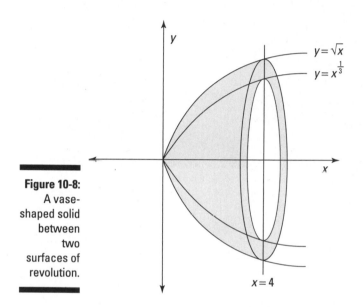

$y = \sqrt{x}$

$y = x^{\frac{1}{3}}$

$x = 4$

Figure 10-8:
A vase-shaped solid between two surfaces of revolution.

This solid looks something like a bowl turned on its side. The outer edge is the solid of revolution around the x-axis for the function $\sqrt{x}$. The inner edge is the solid of revolution around the x-axis for the function $x^{\frac{1}{3}}$.

1. **Find an expression that represents the area of a random cross section of the solid.**

   That is, find the area of a circle with a radius of $\sqrt{x}$ and subtract the area of a circle with a radius of $x^{\frac{1}{3}}$:

   $$A = \pi\left(\sqrt{x}\right)^2 - \pi\left(x^{\frac{1}{3}}\right)^2 = \pi\left(x - x^{\frac{2}{3}}\right)$$

2. **Use this expression to build a definite integral that represents the volume of the solid.**

   The limits of integration this time are 0 and 4:

   $$V = \int_0^4 \pi\left(x - x^{\frac{2}{3}}\right) dx$$

3. Solve the integral:

$$= \pi \int_0^4 \left(x - x^{\frac{2}{3}} \right) dx$$

$$= \pi \left(\int_0^4 x\, dx - \int_0^4 x^{\frac{2}{3}} dx \right)$$

$$= \pi \left(\frac{1}{2} x^2 \Big|_{x=0}^{x=4} - \frac{3}{5} x^{\frac{5}{3}} \Big|_{x=0}^{x=4} \right)$$

Now, evaluate this expression:

$$= \pi \left[\left(\frac{1}{2} 4^2 - 0 \right) - \left(\frac{3}{5} 4^{\frac{5}{3}} - 0 \right) \right]$$

$$= \pi \left(8 - \frac{3}{5} 1{,}024^{\frac{1}{3}} \right)$$

$$\approx 6.1336$$

Here's a problem that brings together everything you've worked with from the meat-slicer method: Find the volume of the solid shown in Figure 10-9. This solid falls between the surface of revolution $y = \ln x$ and the surface of revolution $y = x^{\frac{3}{4}}$, bounded below by $y = 0$ and above by $y = 1$.

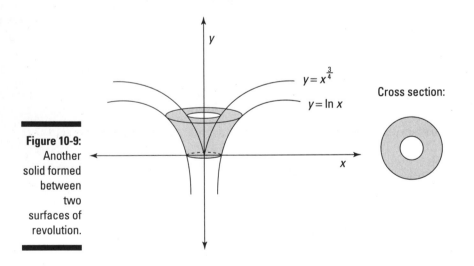

Figure 10-9:
Another
solid formed
between
two
surfaces of
revolution.

The cross section of this solid is shown in the right side of Figure 10-9: a circle with a hole in the middle.

Notice, however, that this cross section is perpendicular to the *y*-axis. To use the meat-slicer method, the cross section must be perpendicular to the *x*-axis. Modify the problem using inverses, as I show you in "Turning a Problem on Its Side":

$$x = \ln y \qquad\qquad x = y^{\frac{3}{4}}$$

$$e^x = y \qquad\qquad x^{\frac{4}{3}} = y$$

The resulting problem is shown in Figure 10-10.

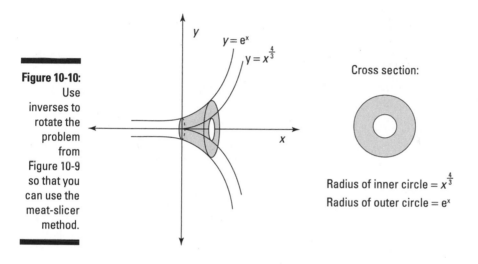

Cross section:

Radius of inner circle $= x^{\frac{4}{3}}$

Radius of outer circle $= e^x$

Now, you can use the meat-slicer method to solve the problem:

1. **Find an expression that represents the area of a random cross section of the solid.**

   That is, find the area of a circle with a radius of e^x and subtract the area of a circle with a radius of $x^{\frac{4}{3}}$. This is just geometry, but I take it slowly so that you can see all the steps. Remember that the area of a circle is πr^2:

   $$A = \text{Area of outer circle} - \text{Area of inner circle}$$

   $$= \pi \left(e^x\right)^2 - \pi \left(x^{\frac{4}{3}}\right)^2$$

   $$= \pi e^{2x} - \pi x^{\frac{8}{3}}$$

2. **Use this expression to build a definite integral that represents the volume of the solid.**

    The limits of integration are 0 and 1:

    $$V = \int_0^1 \left(\pi e^{2x} - \pi x^{\frac{8}{3}} \right) dx$$

3. **Evaluate the integral:**

    $$= \int_0^1 \pi e^{2x} \, dx - \int_0^1 \pi x^{\frac{8}{3}} \, dx$$

    $$= \frac{\pi}{2} e^{2x} \Big|_{x=0}^{x=1} - \frac{3\pi}{11} x^{\frac{11}{3}} \Big|_{x=0}^{x=1}$$

    $$= \left(\frac{\pi}{2} e^2 - \frac{\pi}{2} e^0 \right) - \left(\frac{3\pi}{11}(1)^{\frac{11}{3}} - \frac{3\pi}{11}(0)^{\frac{11}{3}} \right)$$

    $$= \frac{\pi}{2} e^2 - \frac{\pi}{2} - \frac{3\pi}{11}$$

    $$\approx 2.9218$$

    So the volume of this solid is approximately 2.9218 cubic units.

Playing the Shell Game

The *shell method* is an alternative to the meat-slicer method, which I discuss earlier in this chapter. It allows you to measure the volume of a solid by measuring the volume of many concentric surfaces of the volume, called "shells."

Although the shell method works only for solids with circular cross sections, it's ideal for solids of revolution around the *y*-axis, because you don't have to use inverses of functions, as I show you in "Turning a Problem on Its Side." Here's how it works:

1. **Find an expression that represents the area of a random shell of the solid in terms of *x*.**

2. **Use this expression to build a definite integral (in terms of *dx*) that represents the volume of the solid.**

3. **Evaluate this integral.**

As you can see, this method resembles the meat-slicer method. The main difference is that you're measuring the area of shells instead of cross sections.

Peeling and measuring a can of soup

You can use a can of soup — or any other can that has a paper label on it — as a handy visual aid to give you insight into how the shell method works. To start out, go to the pantry and get a can of soup.

Suppose that your can of soup is industrial size, with a radius of 3 inches and a height of 8 inches. You can use the formula for a cylinder to figure out its volume as follows:

$$V = A_b \cdot h = 3^2\pi \cdot 8 = 72\pi$$

Another option is the meat-slicer method, as I show you earlier in this chapter. A third option, which I focus on here, is the shell method.

To understand the shell method, slice the can's paper label vertically, and carefully remove it from the can, as shown in Figure 10-11. (While you're at it, take a moment to read the label so that you're not left with "mystery soup.")

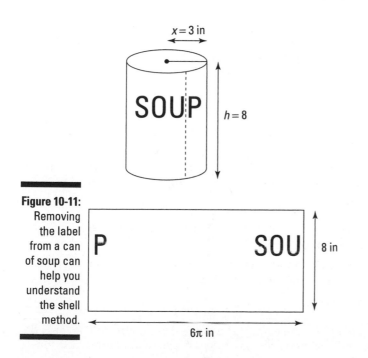

Notice that the label is simply a rectangle. Its shorter side is equal in length to the height of the can (8 inches) and its longer side is equal to the circumference ($2\pi \cdot 3$ inches = 6π inches). So the area of this rectangle is 48 square inches.

Now here's the crucial step: Imagine that the entire can is made up of infinitely many labels wrapped concentrically around each other, all the way to its core. The area of each of these rectangles is:

$$A = 2\pi\, x \cdot 8 = 16\pi\, x$$

The variable x in this case is any possible radius, from 0 (the radius of the circle at the very center of the can) to 3 (the radius of the circle at the outer edge). Here's how you use the shell method, step by step, to find the volume of the can:

1. **Find an expression that represents the area of a random shell of the can (in terms of x):**

   $$A = 2\pi\, x \cdot 8 = 16\pi\, x$$

2. **Use this expression to build a definite integral (in terms of dx) that represents the volume of the can.**

   Remember that with the shell method, you're adding up all the shells from the center (where the radius is 0) to the outer edge (where the radius is 3). So use these numbers as the limits of integration:

   $$V = \int_0^3 16\pi x \, dx$$

3. **Evaluate this integral:**

   $$= 16\pi \cdot \frac{1}{2} x^2 \Big|_{x=0}^{x=3}$$

   $$= 8\pi x^2 \Big|_{x=0}^{x=3}$$

   Now, evaluate this expression:

   $$= 8\pi (3)^2 - 0 = 72\pi$$

   The shell method verifies that the volume of the can is 72π cubic inches.

Using the shell method

One advantage of the shell method over the meat-slicer method comes into play when you're measuring a volume of revolution around the y-axis.

Earlier in this chapter I tell you that the meat-slicer method works best when a solid is on its side — that is, when you can slice it perpendicular to the x-axis. But when the similar cross sections of a solid are perpendicular to the y-axis, you need to use inverses to realign the problem before you can start slicing. (See the earlier section "Turning a Problem on Its Side" for more details.)

This realignment step isn't necessary for the shell method. This makes the shell method ideal for measuring solids of revolution around the y-axis. For example, suppose that you want to measure the volume of the solid shown in Figure 10-12.

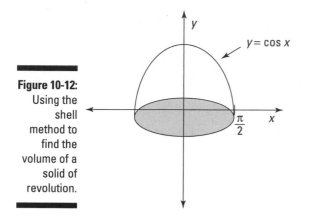

Figure 10-12:
Using the shell method to find the volume of a solid of revolution.

Here's how the shell method can give you a solution without using inverses:

1. **Find an expression that represents the area of a random shell of the solid (in terms of x).**

   Remember that each shell is a rectangle with two different sides: One side is the height of the function at x — that is, cos x. The other is the circumference of the solid at x — that is, $2\pi x$. So, to find the area of a shell, multiply these two numbers together:

   $A = 2\pi x \cos x$

2. **Use this expression to build a definite integral (in terms of dx) that represents the volume of the solid.**

   In this case, remember that you're adding up all the shells from the center (at $x = 0$) to the outer edge (at $x = \frac{\pi}{2}$).

   $$V = \int_0^{\frac{\pi}{2}} 2\pi x \cos x \, dx$$

   $$= 2\pi \int_0^{\frac{\pi}{2}} x \cos x \, dx$$

3. **Evaluate the integral.**

   This integral is pretty easy to solve using integration by parts:

   $$x \sin x + \cos x \Big|_{x=0}^{x=\frac{\pi}{2}}$$

   Now, evaluate this expression:

   $$= \left(\frac{\pi}{2} \sin \frac{\pi}{2} + \cos \frac{\pi}{2} \right) - \left(0 \sin 0 + \cos 0 \right)$$

   $$= \left(\frac{\pi}{2} + 0 \right) - \left(0 + 1 \right)$$

   $$= \frac{\pi}{2} - 1$$

   $$\approx 0.5708$$

   So the volume of the solid is approximately 0.5708 cubic units.

Knowing When and How to Solve 3-D Problems

Because students are so often confused when it comes to solving 3-D calculus problems, here's a final perspective on all the methods in this chapter, and how to choose among them.

First, remember that every problem in this chapter falls into one of these two categories:

✔ Finding the area of a surface of revolution

✔ Finding a volume of a solid

In the first case, use the formula I provide earlier in this chapter, in "Skimming the surface of revolution."

In the second case, remember that the key to measuring the volume of any solid is to slice it up in the direction where it has *similar cross sections* whose area can be measured easily — for example, a circle, a square, or a triangle. So, your first question is whether these similar cross sections are arranged horizontally or vertically.

✔ **Horizontally** means that the solid is already in position for the meat-slicer method. (If it's helpful, imagine slicing a salami in a meat-slicer. The salami must be aligned lying on its side — that is, horizontally — before you can begin slicing.)

✔ **Vertically** means that the solid is standing upright so that the slices are stacked on top of each other.

When the cross sections are arranged horizontally, the meat-slicer method is the easiest way to handle the problem (see "Slicing Your Way to Success" earlier in this chapter).

When the cross sections are arranged vertically, however, your next question is whether these cross sections are circles:

✔ If the cross sections are *not* circles, you must use inverses to flip the solid in the horizontal direction (as I discuss in "Turning a Problem on Its Side").

✔ If they *are* circles, you can either use inverses to flip the solid in the horizontal direction (as I discuss in "Turning a Problem on Its Side") *or* use the shell method (as I discuss in "Playing the Shell Game").

Part IV
Infinite Series

The 5th Wave   By Rich Tennant

"Infinite series has to do with sequences. The sequence I remember best is n+1 over n² equals dazed expression plus pounding headache."

In this part . . .

I introduce the infinite series — that is, the sum of an infinite number of terms. I show you the basics of working with sequences and series, and show you a bunch of ways to determine whether a series is convergent or divergent. You also discover how to use the Taylor series for expressing and evaluating a wide variety of functions.

Chapter 11

Following a Sequence, Winning the Series

*J*ust when you think the semester is winding down, your Calculus II professor introduces a new topic: infinite series.

When you get right down to it, series aren't really all that difficult. After all, a series is just a bunch of numbers added together. Sure, it happens that this bunch is infinite, but addition is just about the easiest math on the planet.

But then again, the last month of the semester is crunch time. You're already anticipating final exams and looking forward to a break from studying. By the time you discover that the prof isn't fooling and really does expect you to know this material, infinite series can lead you down an infinite spiral of despair: Why *this?* Why *now?* Why *me?*

In this chapter, I show you the basics of series. First, you wade into these new waters slowly by examining infinite sequences. When you understand sequences, series make a whole lot more sense. Next, I introduce you to infinite series. I discuss how to express a series in both expanded notation and sigma notation, and then I make sure that you're comfortable with sigma notation. I also show you how every series is related to two sequences.

Next, I introduce you to the all-important topic of convergence and divergence. This concept looms large, so I give you the basics in this chapter and save the more complex information for Chapter 12. Finally, I introduce you to a few important types of series.

Introducing Infinite Sequences

A sequence of numbers is simply a bunch of numbers in a particular order. For example:

1, 4, 9, 16, 25, ...

π, 2π, 3π, 4π, ...

$\frac{1}{2}, \frac{1}{3}, \frac{1}{4}, \frac{1}{5}, ...$

2, 3, 5, 7, 11, 13, ...

2, –2, 2, –2, ...

0, 1, –1, 2 –2, 3, ...

When a sequence goes on forever, it's an *infinite sequence*. Calculus — which focuses on all things infinite — concerns itself predominantly with infinite sequences.

Each number in a sequence is called a *term* of that sequence. So, in the sequence 1, 4, 9, 16, ... , the *first term* is 1, the *second term* is 4, and so forth.

Understanding sequences is an important first step toward understanding series.

Understanding notations for sequences

The simplest notation for defining a sequence is a variable with the subscript *n* surrounded by braces. For example:

$\{a_n\} = \{1, 4, 9, 16, ...\}$

$\{b_n\} = \{1, \frac{1}{2}, \frac{1}{3}, \frac{1}{4}, ...\}$

$\{c_n\} = \{4\pi, 6\pi, 8\pi, 10\pi, ...\}$

You can reference a specific term in the sequence by using the subscript:

$a_1 = 1$ $b_3 = \frac{1}{3}$ $c_6 = 14\pi$

WARNING!

Make sure that you understand the difference between notation with and without braces:

✔ The notation $\{a_n\}$ *with* braces refers to the *entire* sequence.

✔ The notation a_n *without* braces refers to the *n*th term of the sequence.

When defining a sequence, instead of listing the first few terms, you can state a rule based on n. (This is similar to how a function is typically defined.) For example:

$\{a_n\}$, where $a_n = n^2$

$\{b_n\}$, where $b_n = \dfrac{1}{n}$

$\{c_n\}$, where $c_n = 2(n + 1)\pi$

Sometimes, for increased clarity, the notation includes the first few terms plus a rule for finding the nth term of the sequence. For example:

$\{a_n\} = \{1, 4, 9, \dots, n^2, \dots\}$

$\{b_n\} = \{1, \dfrac{1}{2}, \dfrac{1}{3}, \dots, \dfrac{1}{n}, \dots\}$

$\{c_n\} = \{4\pi, 6\pi, 8\pi, \dots, 2(n + 1)\pi, \dots\}$

This notation can be made more concise by appending starting and ending values for n:

$$\{a_n\} = \{n^2\}_{n=1}^{\infty}$$

$$\{b_n\} = \left\{\frac{1}{n}\right\}_{n=1}^{\infty}$$

$$\{c_n\} = \{2n\pi\}_{n=2}^{\infty}$$

This last example points out the fact that the initial value of n doesn't have to be 1, which gives you greater flexibility to define a number series by using a rule.

Don't let the fancy notation for number sequences get to you. When you're faced with a new sequence that's defined by a rule, jot down the first four or five numbers in that sequence. Usually, after you see the pattern, you'll find that a problem is much easier.

Looking at converging and diverging sequences

Every infinite sequence is either convergent or divergent:

- A *convergent* sequence has a limit — that is, it approaches a real number.

- A *divergent* sequence doesn't have a limit.

For example, here's a convergent sequence:

$$\{a_n\} = \{1, \frac{1}{2}, \frac{1}{3}, \frac{1}{4}, \frac{1}{5}, ...\}$$

This sequence approaches 0, so:

$$\lim\{a_n\} = 0$$

Thus, this sequence *converges to 0.*

Here's another convergent sequence:

$$\{b_n\} = \{7, 9, 7\frac{1}{2}, 8\frac{1}{2}, 7\frac{3}{4}, 8\frac{1}{4}, ...\}$$

This time, the sequence approaches 8 from above and below, so:

$$\lim\{b_n\} = 8$$

In many cases, however, a sequence *diverges* — that is, it fails to approach any real number. Divergence can happen in two ways. The most obvious type of divergence occurs when a sequence explodes to infinity or negative infinity — that is, it gets farther and farther away from 0 with every term. Here are a few examples:

–1, –2, –3, –4, –5, –6, –7, ...

ln 1, ln 2, ln 3, ln 4, ln 5, ...

2, 3, 5, 7, 11, 13, 17, ...

In each of these cases, the sequence approaches either ∞ or –∞, so the limit of the sequence *does not exist* (DNE). Therefore, the sequence is divergent.

A second type of divergence occurs when a sequence oscillates between two or more values. For example:

0, 7, 0, 7, 0, 7, 0, 7, ...

1, 1, 2, 1, 2, 3, 1, 2, 3, 4, 1, ...

In these cases, the sequence bounces around indefinitely, never settling in on a value. Again, the limit of the sequence *does not exist,* so the sequence is divergent.

Introducing Infinite Series

In contrast to an infinite sequence (which is an endless list of numbers), an *infinite series* is an endless sum of numbers. You can change any infinite sequence to an infinite series simply by changing the commas to plus signs. For example:

$$1, 2, 3, 4, \ldots \qquad\qquad 1 + 2 + 3 + 4 + \ldots$$

$$1, \frac{1}{2}, \frac{1}{3}, \frac{1}{4}, \ldots \qquad\qquad 1 + \frac{1}{2} + \frac{1}{3} + \frac{1}{4} + \ldots$$

$$1, -1, \frac{1}{2}, -\frac{1}{2}, \frac{1}{4}, -\frac{1}{4}, \ldots \qquad 1 + -1 + \frac{1}{2} + -\frac{1}{2} + \frac{1}{4} + -\frac{1}{4} + \ldots$$

The two principal notations for series are sigma notation and expanded notation. *Sigma notation* provides an explicit rule for generating the series (see Chapter 2 for the basics of sigma notation). *Expanded notation* gives enough of the first few terms of a series so that the pattern generating the series becomes clear.

For example, here are three series defined using both forms of notation:

$$\sum_{n=1}^{\infty} 2n = 2 + 4 + 6 + 8 + \ldots$$

$$\sum_{n=0}^{\infty} \frac{1}{4^n} = 1 + \frac{1}{4} + \frac{1}{16} + \frac{1}{64} + \ldots$$

$$\sum_{n=3}^{\infty} \frac{n}{e^n} = \frac{3}{e^3} + \frac{4}{e^4} + \frac{5}{e^5} + \ldots$$

As you can see, a series can start at any integer.

As with sequences (see "Introducing Infinite Sequences" earlier in this chapter), every series is either convergent or divergent:

✔ A *convergent* series evaluates to a real number.

✔ A *divergent* series doesn't evaluate to a real number.

To get clear on how evaluation of a series connects with convergence and divergence, I give you a few examples. To start out, consider this convergent series:

$$\sum_{n=0}^{\infty} \left(\frac{1}{2}\right)^n = 1 + \frac{1}{2} + \frac{1}{4} + \frac{1}{8} + \ldots$$

Notice that as you add this series from left to right, term by term, the running total is a sequence that approaches 2:

$$1, \frac{3}{2}, \frac{7}{4}, \frac{15}{8}, ...$$

This sequence is called the *sequence of partial sums* for this series. I discuss sequences of partial sums in greater detail later in "Connecting a Series with Its Two Related Sequences."

For now, please remember that the value of a series equals the limit of its sequence of partial sums. In this case, because the limit of the sequence is 2, you can evaluate the series as follows:

$$\sum_{n=0}^{\infty} \left(\frac{1}{2}\right)^n = 2$$

Thus, this series *converges to 2.*

Often, however, a series *diverges* — that is, it doesn't equal any real number. As with sequences, divergence can happen in two ways. The most obvious type of divergence occurs when a series explodes to infinity or negative infinity. For example:

$$\sum_{n=1}^{\infty} -n = -1 + -2 + -3 + -4 + ...$$

This time, watch what happens as you add the series term by term:

$$-1, -3, -6, -10, ...$$

Clearly, this sequence of partial sums diverges to negative infinity, so the series is divergent as well.

A second type of divergence occurs when a series alternates between positive and negative values in such a way that the series never approaches a value. For example:

$$\sum_{n=0}^{\infty} (-1)^n = 1 + -1 + 1 + -1 + ...$$

So, here's the related sequence of partial sums:

$$1, 0, 1, 0, ...$$

In this case, the sequence of partial sums alternates forever between 1 and 0, so it's divergent; therefore, the series is also divergent. This type of series is called, not surprisingly, an *alternating series.* I discuss alternating series in greater depth in Chapter 12.

Convergence and divergence are arguably the most important topics in your final weeks of Calculus II. Many of your exam questions will ask you to determine whether a given series is convergent or divergent.

Later in this chapter, I show you how to decide whether certain important types of series are convergent or divergent. Chapter 12 gives you a ton of handy tools for answering this question more generally. For now, just keep this important idea of convergence and divergence in mind.

Getting Comfy with Sigma Notation

Sigma notation is a compact and handy way to represent series.

Okay — that's the official version of the story. What's also true is that sigma notation can be unclear and intimidating — especially when the professor starts scrawling it all over the blackboard at warp speed while explaining some complex proof. Lots of students get left in the chalk dust (or dry-erase marker fumes).

At the same time, sigma notation is useful and important because it provides a concise way to express series and mathematically manipulate them.

In this section, I give you a bunch of handy tips for working with sigma notation. Some of the uses for these tips become clearer as you continue to study series later in this chapter and in Chapters 12 and 13. For now, just add these tools to your toolbox and use them as needed.

Writing sigma notation in expanded form

When you're working with an unfamiliar series, begin by writing it out using both sigma and expanded notation. This practice is virtually guaranteed to increase your understanding of the series. For example:

$$\sum_{n=1}^{\infty} \frac{2^n}{3n}$$

As it stands, you may not have much insight into what this series looks like, so expand it out:

$$\sum_{n=1}^{\infty} \frac{2^n}{3n} = \frac{2}{3} + \frac{4}{6} + \frac{8}{9} + \frac{16}{12} + \frac{32}{16} + \dots$$

As you spend a bit of time generating this series, it begins to grow less frightening. For one thing, you may notice that in a race between the numerator and denominator, eventually the numerator catches up and pulls ahead. Because the terms eventually grow greater than 1, the series explodes to infinity, so it *diverges*.

Seeing more than one way to use sigma notation

Virtually any series expressed in sigma notation can be rewritten in a slightly altered form. For example:

$$\frac{1}{8} + \frac{1}{16} + \frac{1}{32} + \frac{1}{64} + \dots$$

You can express this series in sigma notation as follows:

$$\sum_{n=3}^{\infty} \left(\frac{1}{2}\right)^n = \frac{1}{8} + \frac{1}{16} + \frac{1}{32} + \frac{1}{64} + \dots$$

Alternatively, you can express the same series in any of the following ways:

$$= \sum_{n=2}^{\infty} \left(\frac{1}{2}\right)^{n+1}$$

$$= \sum_{n=1}^{\infty} \left(\frac{1}{2}\right)^{n+2}$$

$$= \sum_{n=0}^{\infty} \left(\frac{1}{2}\right)^{n+3}$$

Depending on the problem that you're trying to solve, you may find one of these expressions more advantageous than the others — for example, when using the comparison tests that I introduce in Chapter 12. For now, just be sure to keep in mind the flexibility at your disposal when expressing a series in sigma notation.

Discovering the Constant Multiple Rule for series

In Chapter 4, you discover that the Constant Multiple Rule for Integration allows you to simplify an integral by factoring out a constant. This option is also available when you're working with series. Here's the rule:

$$\sum c a_n = c \sum a_n$$

For example:

$$\sum_{n=1}^{\infty} \frac{7}{n^2} = 7 \sum_{n=1}^{\infty} \frac{1}{n^2}$$

To see why this rule works, first expand the series so that you can see what you're working with:

$$\sum_{n=1}^{\infty} \frac{7}{n^2} = 7 + \frac{7}{4} + \frac{7}{9} + \frac{7}{16} + \dots$$

Working with the expanded form, you can factor out a 7 from each term:

$$= 7 \left(1 + \frac{1}{4} + \frac{1}{9} + \frac{1}{16} + \dots \right)$$

Now, express the contents of the parentheses in sigma notation:

$$= 7 \sum_{n=1}^{\infty} \frac{1}{n^2}$$

As if by magic, this procedure demonstrates that the two sigma expressions are equal. But, this magic is really nothing more exotic than your old friend from grade school, the distributive property.

Examining the Sum Rule for series

Here's another handy tool for your growing toolbox of sigma tricks. This rule mirrors the Sum Rule for Integration (see Chapter 4), which allows you to split a sum inside an integral into the sum of two separate integrals. Similarly, you can break a sum inside a series into the sum of two separate series:

$$\Sigma (a_n + b_n) = \Sigma a_n + \Sigma b_n$$

For example:

$$= \sum_{n=1}^{\infty} \frac{n+1}{2^n}$$

A little algebra allows you to split this fraction into two terms:

$$= \sum_{n=1}^{\infty} \left(\frac{n}{2^n} + \frac{1}{2^n} \right)$$

Now, the rule allows you to split this result into two series:

$$= \sum_{n=1}^{\infty} \frac{n}{2^n} + \sum_{n=1}^{\infty} \frac{1}{2^n}$$

This sum of two series is equivalent to the series that you started with. As with the Sum Rule for Integration, expressing a series as a sum of two simpler series tends to make problem-solving easier. Generally speaking, as you proceed onward with series, any trick you can find to simplify a difficult series is a good thing.

Connecting a Series with Its Two Related Sequences

Every series has two related sequences. The distinction between a sequence and a series is as follows:

- A sequence is a *list* of numbers separated by *commas* (for example: 1, 2, 3, ...).
- A series is a *sum* of numbers separated by *plus signs* (for example: 1 + 2 + 3 + ...).

When you see how a series and its two related sequences are distinct but also related, you gain a clearer understanding of how series work.

A series and its defining sequence

The first sequence related to a series is simply the sequence that defines the series in the first place. For example, here are three series written in both sigma notation and expanded notation, each paired with its defining sequence:

$$\sum_{n=1}^{\infty}\left(\frac{1}{2}\right)^n = 1 + \frac{1}{2} + \frac{1}{4} + \frac{1}{8} + \dots \qquad 1, \frac{1}{2}, \frac{1}{4}, \frac{1}{8}, \dots$$

$$\sum_{n=1}^{\infty}\frac{n}{n+1} = \frac{1}{2} + \frac{2}{3} + \frac{3}{4} + \frac{4}{5} + \dots \qquad \frac{1}{2}, \frac{2}{3}, \frac{3}{4}, \frac{4}{5}, \dots$$

$$\sum_{n=1}^{\infty}\frac{1}{n} = 1 + \frac{1}{2} + \frac{1}{3} + \frac{1}{4} + \dots \qquad 1, \frac{1}{2}, \frac{1}{3}, \frac{1}{4}, \dots$$

When a sequence $\{a_n\}$ is already defined, you can use the notation $\Sigma\, a_n$ to refer to the related series starting at $n = 1$. For example, when $\{a_n\} = \frac{1}{n^2}$, $\Sigma\, a_n = 1 + \frac{1}{4} + \frac{1}{9} + \frac{1}{16} + \dots$.

Understanding the distinction between a series and the sequence that defines it is important for two reasons. First, and most basic, you don't want to get the concepts of sequences and series confused. But second, the sequence that defines a series can provide important information about the series. See Chapter 12 to find out about the nth term test, which provides a connection between a series and its defining sequence.

A series and its sequences of partial sums

You can learn a lot about a series by finding the *partial sums* of its first few terms. For example, here's a series that you've seen before:

$$\sum_{n=1}^{\infty}\left(\frac{1}{2}\right)^n = \frac{1}{2} + \frac{1}{4} + \frac{1}{8} + \frac{1}{16} + \cdots$$

And here are the first four partial sums of this series:

$$\sum_{n=1}^{1}\left(\frac{1}{2}\right)^n = \frac{1}{2}$$

$$\sum_{n=1}^{2}\left(\frac{1}{2}\right)^n = \frac{1}{2} + \frac{1}{4} = \frac{3}{4}$$

$$\sum_{n=1}^{3}\left(\frac{1}{2}\right)^n = \frac{1}{2} + \frac{1}{4} + \frac{1}{8} = \frac{7}{8}$$

$$\sum_{n=1}^{4}\left(\frac{1}{2}\right)^n = \frac{1}{2} + \frac{1}{4} + \frac{1}{8} + \frac{1}{16} = \frac{15}{16}$$

You can turn the partial sums for this series into a sequence as follows:

$$\{S_n\} = \left\{\frac{1}{2}, \frac{3}{4}, \frac{7}{8}, \frac{15}{16}, \cdots, \frac{2^n}{2^n - 1}, \cdots\right\}$$

In general, every series $\Sigma\, a_n$ has a related sequence of partial sums $\{S_n\}$. For example, here are a few such pairings:

$$\sum_{n=1}^{\infty}\left(\frac{1}{2}\right)^n = \frac{1}{2} + \frac{1}{4} + \frac{1}{8} + \frac{1}{16} + \cdots \qquad \frac{1}{2}, \frac{3}{4}, \frac{7}{8}, \frac{15}{16}, \cdots$$

$$\sum_{n=1}^{\infty}\frac{n}{n+1} = \frac{1}{2} + \frac{2}{3} + \frac{3}{4} + \frac{4}{5} + \cdots \qquad \frac{1}{2}, \frac{7}{6}, \frac{23}{12}, \frac{163}{60}, \cdots$$

$$\sum_{n=1}^{\infty}\frac{1}{n} = 1 + \frac{1}{2} + \frac{1}{3} + \frac{1}{4}, \cdots \qquad 1, \frac{3}{2}, \frac{11}{6}, \frac{25}{12}, \cdots$$

Every series and its related sequence of partial sums are either *both convergent* or *both divergent*. Moreover, if they're both convergent, both converge to the same number.

This rule should come as no big surprise. After all, a sequence of partial sums simply gives you a running total of where a series is going. Still, this rule can be helpful. For example, suppose that you want to know whether the following sequence is convergent or divergent:

$$1, \frac{3}{2}, \frac{11}{6}, \frac{25}{12}, \frac{137}{60}, \cdots$$

What the heck is this sequence, anyway? Upon deeper examination, however, you discover that it's the sequence of partial sums for every simple series:

$$1$$
$$1 + \frac{1}{2} = \frac{3}{2}$$
$$1 + \frac{1}{2} + \frac{1}{3} = \frac{11}{6}$$
$$1 + \frac{1}{2} + \frac{1}{3} + \frac{1}{4} = \frac{25}{12}$$
$$1 + \frac{1}{2} + \frac{1}{3} + \frac{1}{4} + \frac{1}{5} = \frac{137}{60}$$

This series, called the harmonic series, is divergent, so you can conclude that its sequence of partial sums also diverges.

Recognizing Geometric Series and P-Series

At first glance, many series look strange and unfamiliar. But a few big categories of series belong in the Hall of Fame. When you know how to identify these types of series, you have a big head start on discovering whether they're convergent or divergent. In some cases, you can also find out the exact value of a convergent series without spending all eternity adding numbers.

In this section, I show you how to recognize and work with two common types of series: geometric series and *p*-series.

Getting geometric series

A geometric series is any series of the following form:

$$\sum_{n=0}^{\infty} ar^n = a + ar + ar^2 + ar^3 + \ldots$$

Here are a few examples of geometric series:

$$\sum_{n=0}^{\infty} 2^n = 1 + 2 + 4 + 8 + 16 + \ldots$$

$$\sum_{n=0}^{\infty} \frac{1}{10^n} = 1 + \frac{1}{10} + \frac{1}{100} + \frac{1}{1,000} + \ldots$$

$$\sum_{n=0}^{\infty} \frac{3}{100^n} = 3 + \frac{3}{100} + \frac{3}{10,000} + \frac{3}{1,000,000} + \ldots$$

In the first series, $a = 1$ and $r = 2$. In the second, $a = 1$ and $r = \frac{1}{10}$. And in the third, $a = 3$ and $r = \frac{1}{100}$.

If you're unsure whether a series is geometric, you can test it as follows:

1. **Let a equal the first term of the series.**

2. **Let r equal the second term divided by the first term.**

3. **Check to see whether the series fits the form $a + ar^2 + ar^3 + ar^4 + \ldots$.**

For example, suppose that you want to find out whether the following series is geometric:

$$\frac{8}{5} + \frac{6}{5} + \frac{9}{10} + \frac{27}{40} + \frac{81}{160} + \frac{243}{640} + \ldots$$

Use the procedure I outline as follows:

1. **Let a equal the first term of the series:**

   $$a = \frac{8}{5}$$

2. **Let r equal the second term divided by the first term:**

   $$r = \frac{6}{5} \div \frac{8}{5} = \frac{3}{4}$$

3. **Check to see whether the series fits the form $a + ar^2 + ar^3 + ar^4 + \dots$:**

$$a = \frac{8}{5}$$

$$ar = \frac{8}{5}\left(\frac{3}{4}\right) = \frac{6}{5}$$

$$ar^2 = \frac{6}{5}\left(\frac{3}{4}\right)^2 = \frac{9}{10}$$

$$ar^3 = \frac{9}{10}\left(\frac{3}{4}\right)^3 = \frac{27}{40}$$

As you can see, this series is geometric. To find the limit of a geometric series $a + ar + ar^2 + ar^3 + \dots$, use the following formula:

$$\sum_{n=0}^{\infty} ar^n = \frac{a}{1-r}$$

So, the limit of the series in the previous example is:

$$\sum_{n=0}^{\infty} \frac{8}{5}\left(\frac{3}{4}\right)^n = \frac{\frac{8}{5}}{1 - \frac{3}{4}} = \frac{8}{5} \cdot \frac{1}{4} = \frac{32}{5}$$

When the limit of a series exists, as in this example, the series is called *convergent*. So, you say that this series *converges* to $\frac{32}{5}$.

In some cases, however, the limit of a geometric series does not exist (DNE). In that case, the series is *divergent*. Here's the complete rule that tells you whether a series is convergent or divergent:

For any geometric series $a + ar + ar^2 + ar^3 + \dots$, if r falls in the open set $(-1, 1)$, the series converges to $\frac{a}{1-r}$; otherwise, the series diverges.

An example makes clear why this is so. Look at the following geometric series:

$$1 + \frac{5}{4} + \frac{25}{16} + \frac{125}{64} + \frac{625}{256} + \dots$$

In this case, $a = 1$ and $r = \frac{5}{4}$. Because $r > 1$, each term in the series is greater than the term that precedes it, so the series grows at an ever-accelerating rate.

This series illustrates a simple but important rule of thumb for deciding whether a series is convergent or divergent: A series can be convergent only when its related sequence *converges to zero*. I discuss this important idea (called the *nth-term test*) further in Chapter 12.

Similarly, look at this example:

$$1 + -\frac{5}{4} + \frac{25}{16} + -\frac{125}{64} + \frac{625}{256} + \dots$$

This time, $a = 1$ and $r = -\frac{5}{4}$. Because $r < -1$, the odd terms grow increasingly positive and the even terms grow increasingly negative. So the related sequence of partial sums alternates wildly from the positive to the negative, with each term further from zero than the preceding term.

A series in which alternating terms are positive and negative is called an *alternating series*. I discuss alternating series in greater detail in Chapter 12.

Generally speaking, the geometric series is the only type of series that has a simple formula to calculate its value. So, when a problem asks for the value of a series, try to put it in the form of a geometric series.

For example, suppose that you're asked to calculate the value of this series:

$$\frac{5}{7} + \frac{10}{21} + \frac{20}{63} + \frac{40}{189} + \ldots$$

The fact that you're being asked to calculate the value of the series should tip you off that it's geometric. Use the procedure I outline earlier to find a and r:

$$a = \frac{5}{7}$$

$$r = \frac{10}{21} \div \frac{5}{7} = \frac{2}{3}$$

So here's how to express the series in sigma notation as a geometric series in terms of a and r:

$$\sum_{n=1}^{\infty} \frac{5}{7}\left(\frac{2}{3}\right)^n = \frac{5}{7} + \frac{10}{21} + \frac{20}{63} + \frac{40}{189} + \ldots$$

At this point, you can use the formula for calculating the value of this series:

$$= \frac{a}{1-r} = \frac{\frac{5}{7}}{\left(1 - \frac{2}{3}\right)} = \frac{5}{7} \cdot \frac{1}{3} = \frac{15}{7}$$

Pinpointing p-series

Another important type of series is called the *p-series*. A *p-series* is any series in the following form:

$$\sum_{n=1}^{\infty} \frac{1}{n^p} = 1 + \frac{1}{2^p} + \frac{1}{3^p} + \frac{1}{4^p} + \ldots$$

Here's a common example of a *p-series*, when $p = 2$:

$$\sum_{n=1}^{\infty} \frac{1}{n^2} = 1 + \frac{1}{4} + \frac{1}{9} + \frac{1}{16} + \ldots$$

Here are a few other examples of *p*-series:

$$\sum_{n=1}^{\infty} \frac{1}{n^5} = 1 + \frac{1}{32} + \frac{1}{243} + \frac{1}{1{,}024} + \dots$$

$$\sum_{n=1}^{\infty} \frac{1}{n^{\frac{1}{2}}} = 1 + \frac{1}{\sqrt{2}} + \frac{1}{\sqrt{3}} + \frac{1}{2} + \frac{1}{\sqrt{5}} + \dots$$

$$\sum_{n=1}^{\infty} \frac{1}{n^{-1}} = 1 + 2 + 3 + 4 + \dots$$

Don't confuse *p*-series with geometric series (which I introduce in the previous section). Here's the difference:

- A geometric series has the variable *n* in the exponent — for example, $\Sigma \left(\frac{1}{2}\right)^n$.
- A *p*-series has the variable in the base — for example $\Sigma \frac{1}{n^2}$.

As with geometric series, a simple rule exists for determining whether a *p*-series is convergent or divergent.

A *p*-series converges when *p* > 1 and diverges when *p* ≤ 1.

I give you a proof of this theorem in Chapter 12. In this section, I show you why a few important examples of *p*-series are either convergent or divergent.

Harmonizing with the harmonic series

When *p* = 1, the *p*-series takes the following form:

$$\sum_{n=1}^{\infty} \frac{1}{n} = 1 + \frac{1}{2} + \frac{1}{3} + \frac{1}{4} + \dots$$

This *p*-series is important enough to have its own name: the *harmonic series*. The harmonic series is *divergent*.

Testing p-series when p = 2, p = 3, and p = 4

Here are the *p*-series when *p* equals the first few counting numbers greater than 1:

$$\sum_{n=1}^{\infty} \frac{1}{n^2} = 1 + \frac{1}{4} + \frac{1}{9} + \frac{1}{16} + \dots$$

$$\sum_{n=1}^{\infty} \frac{1}{n^3} = 1 + \frac{1}{8} + \frac{1}{27} + \frac{1}{64} + \dots$$

$$\sum_{n=1}^{\infty} \frac{1}{n^4} = 1 + \frac{1}{16} + \frac{1}{81} + \frac{1}{256} + \dots$$

Because *p* > 1, these series are all *convergent*.

Testing p-series when $p = \frac{1}{2}$

When $p = \frac{1}{2}$, the p-series looks like this:

$$\sum_{n=1}^{\infty} \frac{1}{n^{\frac{1}{2}}} = 1 + \frac{1}{\sqrt{2}} + \frac{1}{\sqrt{3}} + \frac{1}{2} + \frac{1}{\sqrt{5}} + \dots$$

Because $p \leq 1$, this series *diverges*. To see why it diverges, notice that when n is a square number, the nth term equals $\frac{1}{n}$. So this p-series includes every term in the harmonic series plus many more terms. Because the harmonic series is divergent, this series is also divergent.

Chapter 12

Where Is This Going? Testing for Convergence and Divergence

Testing for convergence and divergence is The Main Event in your Calculus II study of series. Recall from Chapter 11 that when a series *converges,* it can be evaluated as a real number. However, when a series *diverges,* it can't be evaluated as a real number, because it either explodes to positive or negative infinity or fails to settle in on a single value.

In Chapter 11, I give you two tests for determining whether specific types of series (geometric series and *p*-series) are convergent or divergent. In this chapter, I give you seven more tests that apply to a much wider range of series.

The first of these is the *n*th-term test, which is sort of a no-brainer. With this under your belt, I move on to two comparison tests: the direct comparison test and the limit comparison test. These tests are what I call one-way tests; they provide an answer only if the series passes the test but not if the series fails it. Next, I introduce three two-way tests, which provide one answer if the series passes the test and the opposite answer if the series fails it. These tests are the integral test, the ratio test, and the root test.

Finally, I introduce you to alternating series, in which terms are alternately positive and negative. I contrast alternating series with positive series, which are the series that you're already familiar with, and I show you how to turn a positive series into an alternating series and vice versa. Then I show you how to prove whether an alternating series is convergent or divergent by using the alternating series test. To finish up, I introduce you to the important concepts of absolute convergence and conditional convergence.

Starting at the Beginning

When testing for convergence or divergence, don't get too hung up on where the series starts. For example:

$$\sum_{n=1,001}^{\infty} \frac{1}{n}$$

This is just a harmonic series with the first 1,000 terms lopped off:

$$= \frac{1}{1,001} + \frac{1}{1,002} + \frac{1}{1,003} + \cdots$$

These fractions may look tiny, but the harmonic series diverges (see Chapter 11), and removing a finite number of terms from the beginning of this series doesn't change this fact.

The lesson here is that, when you're testing for convergence or divergence, what's going on at the beginning of the series is irrelevant. Feel free to lop off the first few billion or so terms of a series if it helps you to prove that the series is convergent or divergent.

Similarly, in most cases you can add on a few terms to a series without changing whether it converges or diverges. For example:

$$\sum_{n=1,000}^{\infty} \frac{1}{n-1}$$

You can start this series anywhere from $n = 2$ to $n = 999$ without changing the fact that it diverges (because it's a harmonic series). Just be careful, because if you try to start the series from $n = 1$, you're adding the term $\frac{1}{0}$, which is a big no-no. However, in most cases you can extend an infinite series without causing problems or changing the convergence or divergence of the series.

Although eliminating terms from the beginning of a series doesn't affect whether the series is convergent or divergent, it *does* affect the sum of a convergent series. For example:

$$\left(\frac{1}{2}\right)^n = 1 + \frac{1}{2} + \frac{1}{4} + \frac{1}{8} + \cdots$$

Lopping off the first few terms of this series — say, 1, $\frac{1}{2}$, and $\frac{1}{4}$ — doesn't change the fact that it's convergent. But it does change the value that the series converges to. For example:

$$\sum_{n=1}^{\infty} \left(\frac{1}{2}\right)^n = \frac{1}{2} + \frac{1}{4} + \frac{1}{8} + \cdots = 1$$

Using the nth-Term Test for Divergence

The nth-term test for divergence is the first test that you need to know. It's easy and it enables you to identify lots of series as divergent.

If the limit of sequence $\{a_n\}$ doesn't equal 0, then the series $\Sigma\, a_n$ is *divergent*.

To show you why this test works, I define a sequence that meets the necessary condition — that is, a sequence that doesn't approach 0:

$$\{a_n\} = \frac{1}{2}, \frac{2}{3}, \frac{3}{4}, \ldots, \frac{n}{n+1} \ldots$$

Notice that the limit of the sequence is 1 rather than 0. So, here's the related series:

$$\sum_{n=1}^{\infty} \frac{n}{n+1} = \frac{1}{2} + \frac{2}{3} + \frac{3}{4} + \ldots$$

Because this series is the sum of an infinite number of terms that are very close to 1, it naturally produces an infinite sum, so it's divergent.

The fact that the limit of a sequence $\{a_n\}$ equals 0 doesn't necessarily imply that the series $\Sigma\, a_n$ is convergent.

For example, the harmonic sequence $1, \frac{1}{2}, \frac{1}{3}, \ldots$ approaches 0, but (as I demonstrate in Chapter 11) the harmonic series $1 + \frac{1}{2} + \frac{1}{3} + \ldots$ is divergent.

When testing for convergence or divergence, always perform the nth-term test first. It's a simple test, and plenty of teachers test for it on exams because it's easy to grade but still catches the unwary student. ***Remember:*** If the defining sequence of a series doesn't approach 0, the series diverges; otherwise, you need to move on to other tests.

Let Me Count the Ways

Tests for convergence or divergence tend to fall into two categories: one-way tests and two-way tests.

One-way tests

A one-way test allows you to draw a conclusion only when a series passes the test, but not when it fails. Typically, *passing the test* means that a given condition has been met.

The nth-term test is a perfect example of a one-way test: If a series *passes* the test — that is, if the limit of its defining sequence doesn't equal 0 — the series is *divergent*. But if the series *fails* the test, you can draw no conclusion.

Later in this chapter, you discover two more one-way tests: the direct comparison test and the limit comparison test.

Two-way tests

A two-way test allows you to draw one conclusion when a series passes the test and the opposite conclusion when a series fails the test. As with a one-way test, *passing the test* means that a given condition has been met. *Failing the test* means that the negation of that condition has been met.

For example, the test for geometric series is a two-way test (see Chapter 11 to find out more about testing geometric series for convergence and divergence). If a series passes the test — that is, if r falls in the open set $(-1, 1)$ — then the series is convergent. And if the series fails the test — that is, if $r \leq -1$ or $r \geq 1$ — then the series is divergent.

Similarly, the test for p-series is also a two-way test (see Chapter 11 for more on this test).

Keep in mind that no test — even a two-way test — is *guaranteed* to give you an answer. Think of each test as a tool. If you run into trouble trying to cut a piece of wood with a hammer, it's not the hammer's fault: You just chose the wrong tool for the job.

Similarly, if you can't find a clever way to demonstrate either the condition or its negation required by a specific test, you're out of luck. In that case, you may need to use a different test that's better suited to the problem.

Later in this chapter, I show you three more two-way tests: the integral test, the ratio test, and the root test.

Using Comparison Tests

Comparison tests allow you to use stuff that you know to find out stuff that you want to know. The stuff that you know is more eloquently called a *benchmark series* — a series whose convergence or divergence you've already proven. The stuff that you want to know is, of course, whether an unfamiliar series converges or diverges.

As with the nth-term test, comparison tests are one-way tests: When a series passes the test, you prove what you've set out to prove (that is, either convergence or divergence). But when a series fails the test, the result of a comparison test in inconclusive.

In this section, I show you two basic comparison tests: the direct comparison test and the limit comparison test.

Getting direct answers with the direct comparison test

You can use the direct comparison test to prove either convergence or divergence, depending on how you set up the test.

To prove that a series converges:

1. **Find a benchmark series that you know converges.**
2. **Show that each term of the series that you're testing is less than or equal to the corresponding term of the benchmark series.**

To prove that a series diverges:

1. **Find a benchmark series that you know diverges.**
2. **Show that each term of the series you're testing is greater than or equal to the corresponding term of the benchmark series.**

For example, suppose that you're asked to determine whether the following series converges or diverges:

$$\text{Benchmark series: } \sum_{n=1}^{\infty} \frac{1}{n^2+1} = \frac{1}{2} + \frac{1}{5} + \frac{1}{10} + \frac{1}{17} + \dots$$

It's hard to tell just by looking at it whether this particular series is convergent or divergent. However, it looks a bit like a p-series with $p = 2$:

$$\sum_{n=1}^{\infty} \frac{1}{n^2} = 1 + \frac{1}{4} + \frac{1}{9} + \frac{1}{16} + \dots$$

You know that this p-series converges (see Chapter 11 if you're not sure why), so use it as your benchmark series. Now, your task is to show that every term in the series that you're testing is less than the corresponding term of the benchmark series:

First term: $\frac{1}{2} < 1$

Second term: $\frac{1}{5} < \frac{1}{4}$

Third term: $\frac{1}{10} < \frac{1}{9}$

This looks good, but to complete the proof formally, here's what you want to show:

nth term: $\frac{1}{n^2+1} \leq \frac{1}{n^2}$

To see that this statement is true, notice that the numerators are the same, but the denominator $(n^2 + 1)$ is greater than n^2. So, the function $\frac{1}{n^2+1}$ is less than $\frac{1}{n^2}$, which means that every term in the test series is less than the corresponding term in the convergent benchmark series. Therefore, both series are convergent.

As another example, suppose that you want to test the following series for convergence or divergence:

$$\sum_{n=1}^{\infty} \frac{3}{n} = 3 + \frac{3}{2} + 1 + \frac{3}{4} + \frac{3}{5} + \dots$$

This time, the series reminds you of the trusty harmonic series, which you know is divergent:

Benchmark series: $\sum_{n=1}^{\infty} \frac{1}{n} = 1 + \frac{1}{2} + \frac{1}{3} + \frac{1}{4} + \frac{1}{5} + \dots$

Using the harmonic series as your benchmark, compare the two series term by term:

First term: $3 > 1$

Second term: $\frac{3}{2} > \frac{1}{2}$

Third term: $1 > \frac{1}{3}$

Again, you have reason to be hopeful, but to complete the proof formally, you want to show the following:

nth term: $\frac{3}{n} \geq \frac{1}{n}$

This time, notice that the denominators are the same, but the numerator 3 is greater than the numerator 1. So the function $\frac{3}{n}$ is greater than $\frac{1}{n}$.

Again, you've shown that every term in the test series is greater than the corresponding term in the divergent benchmark series, so both series are divergent.

As a third example, suppose that you're asked to show whether this series is convergent or divergent:

$$\sum_{n=1}^{\infty} \frac{1}{(n+1)(n+2)} = \frac{1}{6} + \frac{1}{12} + \frac{1}{20} + \frac{1}{30} + \cdots$$

In this case, multiplying out the denominators is a helpful first step:

$$= \sum_{n=1}^{\infty} \frac{1}{n^2 + 3n + 2}$$

Now, the series looks a little like a *p*-series with *p* = 2, so make this your benchmark series:

$$\sum_{n=1}^{\infty} \frac{1}{n^2} = 1 + \frac{1}{4} + \frac{1}{9} + \frac{1}{16} + \cdots$$

The benchmark series converges, so you want to show that every term of the test series is less than the corresponding term of the benchmark. This looks likely because:

First term: $\frac{1}{6} < 1$

Second term: $\frac{1}{12} < \frac{1}{4}$

Third term: $\frac{1}{20} < \frac{1}{9}$

However, to convince the professor, you want to show that *every* term of the test series is less than the corresponding term:

*n*th term: $\frac{1}{n^2 + 3n + 2} \le \frac{1}{n^2}$

As with the first example in this section, the numerators are the same, but the denominator of the test series is greater than that of the benchmark series. So the test series is, indeed, less than the benchmark series, which means that the test series is also convergent.

Testing your limits with the limit comparison test

As with the direct comparison test, the limit comparison test works by choosing a benchmark series whose behavior you know and using it to provide information about a test series whose behavior you don't know.

Here's the limit comparison test: Given a test series $\Sigma\, a_n$ and a benchmark series $\Sigma\, b_n$, find the following limit:

$$\lim_{n \to \infty} \frac{a_n}{b_n}$$

If this limit evaluates as a positive number, then either both series converge or both diverge.

As with the direct comparison test, when the test succeeds, what you learn depends upon what you already know about the benchmark series. If the benchmark series converges, so does the test series. However, if the benchmark series diverges, so does the test series.

Remember, however, that this is a one-way test: If the test fails, you can draw no conclusion about the test series.

The limit comparison test is especially good for testing infinite series based on *rational expressions*. For example, suppose that you want to see whether the following series converges or diverges:

$$\sum_{n=1}^{\infty} \frac{n-5}{n^2+1}$$

When testing an infinite series based on a rational expression, choose a benchmark series that's proportionally similar — that is, whose numerator and denominator differ by the same number of degrees.

In this example, the numerator is a first-degree polynomial and the denominator is a second-degree polynomial (for more on polynomials, see Chapter 2). So the denominator is one degree greater than the numerator. Therefore, I choose a benchmark series that's proportionally similar — the trusty harmonic series:

Benchmark series: $\displaystyle\sum_{n=1}^{\infty} \frac{1}{n}$

Before you begin, take a moment to get clear on what you're testing, and jot it down. In this case, you know that the benchmark series diverges. So, if the test succeeds, you prove that the test series also diverges. (If it fails, however, you're back to square one because this is a one-way test.)

Now, set up the limit (by the way, it doesn't matter which series you put in the numerator and which in the denominator):

$$\lim_{n \to \infty} \frac{\dfrac{n-5}{n^2+1}}{\dfrac{1}{n}}$$

At this point, you just crunch the numbers:

$$\lim_{n \to \infty} \frac{(n-5)n}{n^2+1}$$

$$=\lim_{n \to \infty} \frac{n^2-5n}{n^2+1}$$

Notice at this point that the numerator and denominator are both second-degree polynomials. Now, as you apply L'Hospital's Rule (taking the derivative of both the numerator and denominator), watch what happens:

$$=\lim_{n \to \infty} \frac{2n-5}{2n}$$

$$=\lim_{n \to \infty} \frac{2}{2} = 1$$

As if by magic, the limit evaluates to a positive number, so the test succeeds. Therefore, the test series diverges. Remember, however, that you made this magic happen by choosing a benchmark series in proportion to the test series.

Another example should make this crystal clear. Discover whether this series is convergent or divergent:

$$\lim_{n \to \infty} \frac{n^3-2}{4n^5-n^3-2}$$

When you see that this series is based on a rational expression, you immediately think of the limit comparison test. Because the denominator is two degrees higher than the numerator, choose a benchmark series with the same property:

Benchmark series: $\sum_{n=1}^{\infty} \frac{1}{n^2}$

Before you begin, jot down the following: The benchmark converges, so if the test succeeds, the test series also converges. Next, set up your limit:

$$\lim_{n \to \infty} \frac{\dfrac{n^3-2}{4n^5-n^3-2}}{\dfrac{1}{n^2}}$$

Now, just solve the limit:

$$=\lim_{n \to \infty} \frac{(n^3-2)n^2}{4n^5-n^3-2}$$

$$=\lim_{n \to \infty} \frac{n^5-2n^2}{4n^5-n^3-2}$$

Again, the numerator and denominator have the same degree, so you're on the right track. Now, solving the limit is just a matter of grinding through a few iterations of L'Hospital's Rule:

$$\lim_{n \to \infty} \frac{5n^4 - 4n}{20n^4 - 3n^2}$$

$$= \lim_{n \to \infty} \frac{20n^3 - 4}{80n^3 - 6n}$$

$$= \lim_{n = \infty} \frac{60n^2}{240n^2 - 6}$$

$$= \lim_{n \to \infty} \frac{120n}{480n}$$

$$= \lim_{n \to \infty} \frac{120}{480} = \frac{1}{4}$$

The test succeeds, so the test series converges. And again, the success of the test was prearranged because you chose a benchmark series in proportion to the test series.

Two-Way Tests for Convergence and Divergence

Earlier in this chapter, I give you a variety of tests for convergence or divergence that work in one direction at a time. That is, passing the test gives you an answer, but failing it provides no information.

The tests in this section all have one important feature in common: Regardless of whether the series passes or fails, whenever the test gives you an answer, that answer *always* tells you whether the series is convergent or divergent.

Integrating a solution with the integral test

Just when you thought that you wouldn't have to think about integration again until two days before your final exam, here it is again. The good news is that the integral test gives you a two-way test for convergence or divergence.

REMEMBER

Here's the integral test:

For any series of the form

$$\sum_{x=a}^{\infty} f(x)$$

consider its associated integral

$$\int_{a}^{\infty} f(x)\,dx$$

If this integral converges, the series also converges; however, if this integral diverges, the series also diverges.

In most cases, you use this test to find out whether a series converges or diverges by testing its associated integral. Of course, changing the series to an integral makes all the integration tricks that you already know and love available to you.

For example, here's how to use the integral test to show that the harmonic series is divergent. First, the series:

$$\sum_{x=1}^{\infty} \frac{1}{x} = 1 + \frac{1}{2} + \frac{1}{3} + \frac{1}{4} + \dots$$

The integral test tells you that this series converges or diverges depending upon whether the following definite integral converges or diverges:

$$\int_{1}^{\infty} \frac{1}{x}\,dx$$

To evaluate this improper integral, express it as a limit, as I show you in Chapter 9:

$$= \lim_{c \to \infty} \int_{1}^{c} \frac{1}{x}\,dx$$

This is simple to integrate and evaluate:

$$= \lim_{c \to \infty} \left(\ln x \Big|_{x=1}^{x=c} \right)$$

$$= \lim_{c \to \infty} \ln c - \ln 1$$

$$\lim_{c \to \infty} \ln c - 0 = \infty$$

Because the limit explodes to infinity, the integral doesn't exist. Therefore, the integral test tells you that the harmonic series is divergent.

As another example, suppose that you want to discover whether the following series is convergent or divergent:

$$\sum_{n=2}^{\infty} \frac{1}{n \ln n}$$

Notice that this series starts at $n = 2$, because $n = 1$ would produce the term $\frac{1}{0}$. To use the integral test, transform the sum into this definite integral, using 2 as the lower limit of integration:

$$\int_{2}^{\infty} \frac{1}{x \ln x} \, dx$$

Again, rewrite this improper integral as the limit of an integral (see Chapter 9):

$$\lim_{c \to \infty} \int_{2}^{c} \frac{1}{x \ln x} \, dx$$

To solve the integral, use the following variable substitution:

$$u = \ln x$$
$$du = \frac{1}{x} dx$$

So you can rewrite the integral as follows:

$$\lim_{c \to \infty} \int_{\ln 2}^{\ln c} \frac{1}{u} \, du$$

Note that as the variable changes from x to u, the limits of integration change from 2 and c to ln 2 and ln c. This change arises when I plug the value $x = 2$ into the equation $u = \ln x$, so $u = \ln 2$. (For more on using variable substitution to evaluate definite integrals, see Chapter 5.)

At this point, you can evaluate the integral:

$$\lim_{c \to \infty} \left(\ln u \Big|_{u = \ln 2}^{u = \ln c} \right)$$

$$\lim_{c \to \infty} \ln(\ln c) - \ln(\ln 2) = \infty$$

You can see without much effort that as c approaches infinity, so does ln c, and the rest of the expression doesn't affect this. Therefore, the series that you're testing is divergent.

Rationally solving problems with the ratio test

The ratio test is especially good for handling series that include factorials. Recall that the factorial of a counting number, represented by the symbol !, is that number multiplied by every counting number less than itself. For example:

$$5! = 5 \cdot 4 \cdot 3 \cdot 2 \cdot 1 = 120$$

Flip to Chapter 2 for some handy tips on factorials that may help you in this section.

To use the ratio test, take the limit (as n approaches ∞) of the $(n + 1)$th term divided by the nth term of the series:

$$\lim_{n \to \infty} \frac{a_{n+1}}{a_n + 1}$$

At the risk of destroying all the trust that you and I have built between us over these pages, I must confess that there are not two, but *three* possible outcomes to the ratio test:

- ✔ If this limit is less than 1, the series converges.
- ✔ If this limit is greater than 1, the series diverges.
- ✔ If this limit equals 1, the test is inconclusive.

But I'm sticking to my guns and calling this a two-way test, because — depending on the outcome — it can potentially prove either convergence or divergence.

For example, suppose that you want to find out whether the following series is convergent or divergent:

$$\sum_{n=1}^{\infty} \frac{2^n}{n!}$$

Before you begin, expand the series so that you can get an idea of what you're working with. I do this in two steps to make sure that the arithmetic is correct:

$$= \frac{2}{1} + \frac{2 \cdot 2}{2 \cdot 1} + \frac{2 \cdot 2 \cdot 2}{3 \cdot 2 \cdot 1} + \frac{2 \cdot 2 \cdot 2 \cdot 2}{4 \cdot 3 \cdot 2 \cdot 1} + \ldots$$

$$= 2 + 2 + \frac{4}{3} + \frac{2}{3} + \frac{4}{15} + \ldots$$

To find out whether this series converges or diverges, set up the following limit:

$$\lim_{n \to \infty} \frac{\dfrac{2^{n+1}}{(n+1)!}}{\dfrac{2^n}{n!}}$$

As you can see, I place the function that defines the series in the denominator. Then I rewrite this function, substituting $n + 1$ for n, and I place the result in the numerator. Now, evaluate the limit:

$$= \lim_{n \to \infty} \frac{(2^{n+1})(n!)}{(n+1)!(2^n)}$$

At this point, to see why the ratio test works so well for exponents and factorials, factor out a 2 from 2^{n+1} and an $n + 1$ from $(n + 1)!$:

$$= \lim_{n \to \infty} \frac{2(2^n)(n!)}{(n+1)(n!)(2^n)}$$

This trick allows you to simplify the limit greatly:

$$= \lim_{n \to \infty} \frac{2}{n+1} = 0 < 1$$

Because the limit is less than 1, the series converges.

Rooting out answers with the root test

The root test works best with series that have powers of n in both the numerator and denominator.

To use the root test, take the limit (as n approaches ∞) of the nth root of the nth term of the series:

$$\lim_{n \to \infty} \sqrt[n]{a_n}$$

As with the ratio test, even though I call this a two-way test, there are really three possible outcomes:

- ✔ If the limit is less than 1, the series converges.
- ✔ If the limit is greater than 1, the series diverges.
- ✔ If the limit equals 1, the test is inconclusive.

For example, suppose that you want to decide whether the following series is convergent or divergent:

$$\sum_{n=1}^{\infty} \frac{(\ln n)^n}{n^n}$$

This would be a very hairy problem to try to solve using the ratio test. To use the root test, take the limit of the nth root of the nth term:

$$\lim_{n \to \infty} \sqrt[n]{\frac{(\ln n)^n}{n^n}}$$

At first glance, this expression looks worse than what you started with. But it begins to look better when you separate the numerator and denominator into two roots:

$$= \lim_{n \to \infty} \frac{\sqrt[n]{(\ln n)^n}}{\sqrt[n]{n^n}}$$

Now, a lot of cancellation is possible:

$$= \lim_{n \to \infty} \frac{\sqrt[n]{\ln n}}{n}$$

Suddenly, the problem doesn't look so bad. The numerator and denominator both approach ∞, so apply L'Hospital's Rule:

$$= \lim_{n \to \infty} \frac{1}{n} = 0 < 1$$

Because the limit is less than 1, the series is convergent.

Alternating Series

Each of the series that I discuss earlier in this chapter (and most of those in Chapter 11) have one thing in common: Every term in the series is positive. So, each of these series is a *positive series*. In contrast, a series that has infinitely many positive and infinitely many negative terms is called an *alternating series*.

Most alternating series flip back and forth between positive and negative terms so that every odd-numbered term is positive and every even-numbered term is negative, or vice versa. This feature adds another spin onto the whole question of convergence and divergence. In this section, I show you what you need to know about alternating series.

Eyeballing two forms of the basic alternating series

The most basic alternating series comes in two forms. In the first form, the odd-numbered terms are negated; in the second, the even-numbered terms are negated.

Without further ado, here's the first form of the basic alternating series:

$$\sum_{n=1}^{\infty}(-1)^{n} = -1 + 1 - 1 + 1 - \ldots$$

As you can see, in this series the odd terms are all negated. And here's the second form, whose even terms are negated:

$$\sum_{n=1}^{\infty}(-1)^{n-1} = 1 - 1 + 1 - 1 + \ldots$$

Obviously, in whichever form it takes, the basic alternating series is divergent because it never converges on a single sum but instead jumps back and forth between two sums for all eternity. Although the functions that produce these basic alternating series aren't of much interest by themselves, they get interesting when they're multiplied by an infinite series.

Making new series from old ones

You can turn any positive series into an alternating series by multiplying the series by $(-1)^{n}$ or $(-1)^{n-1}$. For example, here's an old friend, the harmonic series:

$$\sum_{n=1}^{\infty}\frac{1}{n} = 1 + \frac{1}{2} + \frac{1}{3} + \frac{1}{4} + \ldots$$

To negate the odd terms, multiply by $(-1)^{n}$:

$$\sum_{n=1}^{\infty}(-1)^{n}\frac{1}{n} = -1 + \frac{1}{2} - \frac{1}{3} + \frac{1}{4} - \ldots$$

To negate the even terms, multiply by $(-1)^{n-1}$:

$$\sum_{n=1}^{\infty}(-1)^{n-1}\frac{1}{n} = 1 - \frac{1}{2} + \frac{1}{3} - \frac{1}{4} + \ldots$$

Alternating series based on convergent positive series

If you know that a positive series converges, any alternating series based on this series also converges. This simple rule allows you to list a ton of convergent alternating series. For example:

$$\sum_{n=0}^{\infty}(-1)^n\left(\frac{1}{2}\right)^n = 1 - \frac{1}{2} + \frac{1}{4} - \frac{1}{8} + \dots$$

$$\sum_{n=1}^{\infty}(-1)^{n-1}\frac{1}{n^2} = 1 - \frac{1}{4} + \frac{1}{9} - \frac{1}{16} + \dots$$

$$\sum_{n=1}^{\infty}(-1)^{n-1}\frac{2^n}{n!} = 2 - 2 + \frac{4}{3} - \frac{2}{3} + \frac{4}{15} - \dots$$

The first series is an alternating version of a geometric series with $r = \frac{1}{2}$. The second is an alternating variation on the familiar p-series with $p = 2$. The third is an alternating series based on a series that I introduce in the earlier section "Rationally solving problems with the ratio test." In each case, the non-alternating version of the series is convergent, so the alternating series is also convergent.

I can show you an easy way to see why this rule works. As an example, I use the first series of the three I just gave you. The value of the positive version of this series is simple to compute by using the formula from Chapter 11:

$$\sum_{n=0}^{\infty}\left(\frac{1}{2}\right)^n = 1 + \frac{1}{2} + \frac{1}{4} + \frac{1}{8} + \dots = 2$$

Similarly, if you negate all the terms, the value is just as simple to compute:

$$\sum_{n=0}^{\infty}-\left(\frac{1}{2}\right)^n = -1 - \frac{1}{2} - \frac{1}{4} - \frac{1}{8} - \dots = -2$$

So, if some terms are positive and others are negative, the value of the resulting series must fall someplace between –2 and 2; therefore the series converges.

Using the alternating series test

As I discuss in the previous section, when you know that a positive series is convergent, you can assume that any alternating series based on that series is also convergent. In contrast, some divergent positive series become convergent when transformed into alternating series.

Fortunately, I can give you a simple test to decide whether an alternating series is convergent or divergent.

An alternating series converges if these two conditions are met:

1. **Its defining sequence converges to zero — that is, it passes the nth-term test.**

2. **Its terms are non-increasing (ignoring minus signs) — that is, each term is less than or equal to the term before it.**

These conditions are fairly easy to test for, making the alternating series test one of the easiest tests in this chapter. For example, here are three alternating series:

$$\sum_{n=1}^{\infty} (-1)^{n-1} \frac{1}{n} = 1 - \frac{1}{2} + \frac{1}{3} - \frac{1}{4} + \dots$$

$$\sum_{n=1}^{\infty} (-1)^{n-1} \frac{1}{\sqrt{n}} = 1 - \frac{1}{\sqrt{2}} + \frac{1}{\sqrt{3}} - \frac{1}{2} + \dots$$

$$\sum_{n=2}^{\infty} (-1)^{n} \frac{1}{n\ln n} = \frac{1}{2\ln 2} - \frac{1}{3\ln 3} + \frac{1}{4\ln 4} - \frac{1}{5\ln 5} + \dots$$

Just by eyeballing them, you can see that each of them meets both criteria of the alternating series test, so they're all convergent. Notice, too, that in each case, the positive version of the same series is divergent. This underscores an important point: When a positive series is convergent, an alternating series based on it is also necessarily convergent; but when a positive series is divergent, an alternating series based on it may be either convergent or divergent.

Technically speaking, the alternating series test is a one-way test: If the series passes the test — that is, if both conditions hold — the series is convergent. However, if the series fails the test — that is, if either condition isn't met — you can draw no conclusion.

In practice, however — and I'm going out on a thin mathematical limb here — I'd say that when a series fails the alternating series test, you have strong circumstantial evidence that the series is divergent.

Why do I say this? First of all, notice that the first condition is the good old-fashioned nth-term test. If any series fails this test, you can just chuck it on the divergent pile and get on with the rest of your day.

Second, it's rare when a series — *any series* — meets the first condition but fails to meet the second condition. Sure, it happens, but you really have to hunt around to find a series like that. And even when you find one, the series usually settles down into an ever-decreasing pattern fairly quickly.

For example, take a look at the following alternating series:

$$\sum_{n=2}^{\infty}(-1)^{n-1}\frac{n^2}{2^n}=\frac{1}{2}-1+\frac{9}{8}-1+\frac{25}{32}-\frac{9}{16}+\frac{49}{128}-\cdots$$

Clearly, this series passes the first condition of the alternating series test — the nth-term test — because the denominator explodes to infinity at a much faster rate than the numerator.

What about the second condition? Well, the first three terms are increasing (disregarding sign), but beyond these terms the series settles into an ever-decreasing pattern. So, you can chop off the first few terms and express the same series in a slightly different way:

$$=\frac{1}{2}-1+\frac{9}{8}-1+\sum_{n=5}^{\infty}(-1)^{n-1}\frac{n^2}{2^n}$$

This version of the series passes the alternating series test with flying colors, so it's convergent. Obviously, adding a few constants to this series doesn't make it divergent, so the original series is also convergent.

So, when you're testing an alternating series, here's what you do:

1. **Test for the first condition — that is, apply the nth-term test.**

   If the series fails, it's divergent, so you're done.

2. **If the series passes the nth-term test, test for the second condition — that is, see whether its terms *eventually* settle into a constantly-decreasing pattern (ignoring their sign, of course).**

In most cases, you'll find that a series that meets the first condition also meets the second, which means that the series is convergent.

In the rare cases when an alternating series meets the first condition of the alternating series test but doesn't meet the second condition, you can draw no conclusion about whether that series converges or diverges.

These cases really are rare, but I show you one so that you know what to do in case your professor decides to get cute on an exam:

$$-\frac{1}{10}+\frac{1}{9}-\frac{1}{100}+\frac{1}{99}-\frac{1}{1,000}+\frac{1}{999}-\cdots$$

$$-\frac{1}{10}+\frac{1}{2}-\frac{1}{100}+\frac{1}{3}-\frac{1}{1,000}+\frac{1}{4}-\cdots$$

Both of these series meet the first criteria of the alternating series test but fail to meet the second, so you can draw no conclusion based upon this test. In fact, the first series is convergent and the second is divergent. Spend a little time studying them and I believe that you'll see why. (***Hint:*** Try to break each series apart into two separate series.)

Understanding absolute and conditional convergence

In the previous two sections, I demonstrate this important fact: When a positive series is convergent, an alternating series based on it is also necessarily convergent; but when a positive series is divergent, an alternating series based on it may be either convergent or divergent.

So, for any alternating series, you have three possibilities:

✔ An alternating series is convergent, and the positive version of that series is also convergent.

✔ An alternating series is convergent, but the positive version of that series is divergent.

✔ An alternating series is divergent, so the positive version of that series must also be divergent.

The existence of three possibilities for alternating series makes a new concept necessary: the distinction between *absolute convergence* and *conditional convergence*.

Table 12-1 tells you when an alternating series is absolutely convergent, conditionally convergent, or divergent.

Table 12-1	Understanding Absolute and Conditional Convergence of Alternating Series	
An Alternating Series Is:	*When That Series Is:*	*And Its Related Positive Series Is:*
Absolutely Convergent	Convergent	Convergent
Conditionally Convergent	Convergent	Divergent
Divergent	Divergent	Divergent

Here are a few examples of alternating series that are absolutely convergent:

$$\sum_{n=0}^{\infty} (-1)^n \left(\frac{1}{2}\right)^n = 1 - \frac{1}{2} + \frac{1}{4} - \frac{1}{8} + \dots$$

$$\sum_{n=1}^{\infty} (-1)^{n-1} \frac{1}{n^2} = 1 - \frac{1}{4} + \frac{1}{9} - \frac{1}{16} + \dots$$

$$\sum_{n=1}^{\infty} (-1)^{n-1} \frac{2^n}{n!} = 2 - 2 + \frac{4}{3} - \frac{2}{3} + \frac{4}{15} - \dots$$

I pulled these three examples from "Alternating series based on convergent positive series" earlier in this chapter. In each case, the positive version of the series is convergent, so the related alternating series must be convergent as well. Taken together, these two facts mean that each series converges absolutely.

And here are a few examples of alternating series that are conditionally convergent:

$$\sum_{n=1}^{\infty}(-1)^{n-1}\frac{1}{n} = 1 - \frac{1}{2} + \frac{1}{3} - \frac{1}{4} + \dots$$

$$\sum_{n=1}^{\infty}(-1)^{n-1}\frac{1}{\sqrt{n}} = 1 - \frac{1}{\sqrt{2}} + \frac{1}{\sqrt{3}} - \frac{1}{2} + \dots$$

$$\sum_{n=2}^{\infty}(-1)^{n}\frac{1}{n\ln n} = \frac{1}{2\ln2} - \frac{1}{3\ln3} + \frac{1}{4\ln4} - \frac{1}{5\ln5} + \dots$$

I pulled these examples from "Using the alternating series test" earlier in this chapter. In each case, the positive version of the series diverges, but the alternating series converges (by the alternating series test). So each of these series converges conditionally.

Finally, here are a couple of examples of alternating series that are divergent:

$$\sum_{n=1}^{\infty}(-1)^{n-1}n = 1 - 2 + 3 - 4 + \dots$$

$$\sum_{n=1}^{\infty}(-1)^{n-1}\frac{n}{n+1} = \frac{1}{2} - \frac{2}{3} + \frac{3}{4} - \frac{4}{5} + \dots$$

$$\frac{-1}{10} + \frac{1}{2} - \frac{1}{100} + \frac{1}{3} - \frac{1}{1,000} + \frac{1}{4} - \dots$$

As you can see, the first two series fail the *n*th-term test, which is also the first condition of the alternating series test, so these two series diverge. As for the third series, it's basically a divergent harmonic series *minus* a convergent geometric series — that is, a divergent series with a finite number subtracted from it — so the entire series diverges.

Testing alternating series

Suppose that somebody (like your professor) hands you an alternating series that you've never seen before and asks you to find out whether it's absolutely convergent, conditionally convergent, or divergent. Here's what you do:

1. **Apply the alternating series test.**

   In most cases, this test tells you whether the alternating series is convergent or divergent:

   a. If it's divergent, you're done! (The alternating series is divergent.)

   b. If it's convergent, the series is either absolutely convergent or conditionally convergent. Proceed to Step 2.

   c. If the alternating series test is inconclusive, you can't rule any option out. Proceed to Step 2.

2. **Rewrite the alternating series as a positive series by:**

   a. Removing $(-1)^n$ or $(-1)^{n-1}$ when you're working with sigma notation.

   b. Changing the minus signs to plus signs when you're working with expanded notation.

3. **Test this positive series for convergence or divergence by using any of the tests in this chapter or Chapter 11:**

   a. If the positive series is convergent, the alternating series is absolutely convergent.

   b. If the positive series is divergent *and* the alternating series is convergent, the alternating series is conditionally convergent.

   c. If the positive series is divergent *but* the alternating series test is inconclusive, the series is either conditionally convergent or divergent, but you still can't tell which.

In most cases, you're not going to get through all these steps and still have a doubt about the series. In the unlikely event that you do find yourself in this position, see whether you can break the alternating series into two separate series — one with positive terms and the other with negative terms — and study these two series for whatever clues you can.

Chapter 13

Dressing up Functions with the Taylor Series

*T*he infinite series known as the Taylor series is one of the most brilliant mathematical achievements that you'll ever come across. It's also quite a lot to get your head around. Although many calculus books tend to throw you in the deep end with the Taylor series, I prefer to take you by the hand and help you wade in slowly.

The Taylor series is a specific form of the power series. In turn, it's helpful to think of a power series as a polynomial with an infinite number of terms. So, in this chapter, I begin with a discussion of polynomials. I contrast polynomials with other elementary functions, pointing out a few reasons why mathematicians like polynomials so much (often, to the exclusion of their families and friends).

Then I move on to power series, showing you how to discover when a power series converges and diverges. I also discuss the interval of convergence for a power series, which is the set of *x* values for which that series converges. After that, I introduce you to the Maclaurin series — a simplified, but powerful, version of the Taylor series.

Finally, the main event: the Taylor series. First, I show you how to use the Taylor series to evaluate other functions; you'll definitely need that for your final exam. I introduce you to the Taylor remainder term, which allows you to find the margin of error when making an approximation. To finish up the chapter, I show you why the Taylor series works, which helps to make sense of the series, but may not be strictly necessary for passing an exam.

Elementary Functions

Elementary functions are those familiar functions that you work with all the time in calculus. They include:

- Addition, subtraction, multiplication, and division
- Powers and roots
- Exponential functions and logarithms (usually, the natural log)
- Trig and inverse trig functions
- All combinations and compositions of these functions

In this section, I discuss some of the difficulties of working with elementary functions. In contrast, I show you why a small subset of elementary functions — the polynomials — is much easier to work with. To finish up, I consider the advantages of expressing elementary functions as polynomials when possible.

Knowing two drawbacks of elementary functions

The set of elementary functions is closed under the operation of differentiation. That is, when you differentiate an elementary function, the result is always another elementary function.

Unfortunately, this set isn't closed under the operation of integration. For example, here's an integral that can't be evaluated as an elementary function:

$$\int e^{x^2} dx$$

So, even though the set of elementary functions is large and complex enough to confuse most math students, for you — the calculus guru — it's a rather small pool.

Another problem with elementary functions is that many of them are difficult to evaluate for a given value of x. Even the simple function $\sin x$ isn't so simple to evaluate because (except for 0) every integer input value results in an irrational output for the function. For example, what's the value of sin 3?

Appreciating why polynomials are so friendly

In contrast to other elementary functions, polynomials are just about the friendliest functions around. Here are just a few reasons why:

- Polynomials are easy to integrate (see Chapter 4 to see how to compute the integral of every polynomial).
- Polynomials are easy to evaluate for any value of x.
- Polynomials are infinitely differentiable — that is, you can calculate the value of the first derivative, second derivative, third derivative, and so on, infinitely.

Representing elementary functions as polynomials

In Part II, I show you a set of tricks for computing and integrating elementary functions. Many of these tricks work by taking a function whose integral can't be computed as such and tweaking it into a more friendly form.

For example, using the substitution $u = \sin x$, you can turn the integral on the left into the one on the right:

$$\int \sin^3 x \cos x \ dx = \int u^3 \, du$$

In this case, you're able to turn the product of two trig functions into a polynomial, which is much simpler to work with and easy to integrate.

Representing elementary functions as series

The tactic of expressing complicated functions as polynomials (and other simple functions) motivates much of the study of infinite series.

Although series may seem difficult to work with — and, admittedly, they do pose their own specific set of challenges — they have two great advantages that make them useful for integration:

✔ First, an infinite series breaks easily into terms. So in most cases, you can use the Sum Rule to break a series into separate terms and evaluate these terms individually.

✔ Second, series tend to be built from a recognizable pattern. So, if you can figure out how to integrate one term, you can usually generalize this method to integrate every term in the series.

Specifically, power series include many of the features that make polynomials easy to work with. I discuss power series in the next section.

Power Series: Polynomials on Steroids

In Chapter 11, I introduce the geometric series:

$$\sum_{n=0}^{\infty} ax^n = a + ax + ax^2 + ax^3 + \ldots$$

I also show you a simple formula to figure out whether the geometric series converges or diverges.

The geometric series is a simplified form of a larger set of series called the power series.

A *power series* is any series of the following form:

$$\sum_{n=0}^{\infty} c_n x^n = c_0 + c_1 x + c_2 x^2 + c_3 x^3 + \ldots$$

Notice how the power series differs from the geometric series:

✔ In a geometric series, every term has the same coefficient.

✔ In a power series, the coefficients may be different — usually according to a rule that's specified in the sigma notation.

Here are a few examples of power series:

$$\sum_{n=0}^{\infty} nx^n = x + 2x^2 + 3x^3 + 4x^4 + \ldots$$

$$\sum_{n=0}^{\infty} \frac{1}{2^{n+2}} x^n = \frac{1}{4} + \frac{1}{8} x + \frac{1}{16} x^2 + \frac{1}{32} x^3 + \ldots$$

$$\sum_{n=0}^{\infty} (-1)^n \frac{1}{(2n)!} x^{2n} = 1 - \frac{1}{2!} x^2 + \frac{1}{4!} x^4 - \frac{1}{6!} x^6 + \ldots$$

You can think of a power series as a polynomial with an infinite number of terms. For this reason, many useful features of polynomials (which I describe earlier in this chapter) carry over to power series.

The most general form of the power series is as follows:

$$\sum_{n=0}^{\infty} c_n(x-a)^n = c_0 + c_1(x-a) + c_2(x-a)^2 + c_3(x-a)^3 + \ldots$$

This form is for a power series that's centered at a. Notice that when $a = 0$, this form collapses to the simpler version that I introduce earlier in this section. So a power series in this form is centered at 0.

Integrating power series

In Chapter 4, I show you a three-step process for integrating polynomials. Because power series resemble polynomials, they're simple to integrate by using the same basic process:

1. **Use the Sum Rule to integrate the series term by term.**

2. **Use the Constant Multiple Rule to move each coefficient outside its respective integral.**

3. **Use the Power Rule to evaluate each integral.**

For example, take a look at the following integral:

$$\int \sum_{n=0}^{\infty} \frac{1}{2^{n+2}} x^n \, dx$$

At first glance, this integral of a series may look scary. But to give it a chance to show its softer side, I expand the series out as follows:

$$= \int \left(\frac{1}{4} + \frac{1}{8}x + \frac{1}{16}x^2 + \frac{1}{32}x^3 + \ldots \right) dx$$

Now you can apply the three steps for integrating polynomials to evaluate this integral:

1. **Use the Sum Rule to integrate the series term by term:**

$$= \int \frac{1}{4} \, dx + \int \frac{1}{8}x \, dx + \int \frac{1}{16}x^2 \, dx + \int \frac{1}{32}x^3 \, dx + \ldots$$

2. **Use the Constant Multiple Rule to move each coefficient outside its respective integral:**

$$= \frac{1}{4} \int dx + \frac{1}{8} \int x \, dx + \frac{1}{16} \int x^2 \, dx + \frac{1}{32} \int x^3 \, dx + \ldots$$

3. Use the Power Rule to evaluate each integral:

$$= \frac{1}{4}x + \frac{1}{16}x^2 + \frac{1}{48}x^3 + \frac{1}{128}x^4 + \dots$$

Notice that this result is another power series, which you can turn back into sigma notation:

$$= \sum_{n=0}^{\infty} \frac{1}{(n+1)\,2^{n+2}}\,x^{n+1}$$

Understanding the interval of convergence

As with geometric series and p-series (which I discuss in Chapter 11), an advantage to power series is that they converge or diverge according to a well-understood pattern.

Unlike these simpler series, however, a power series often converges or diverges based on its x value. This leads to a new concept when dealing with power series: the interval of convergence.

The *interval of convergence* for a power series is the set of x values for which that series converges.

The interval of convergence is never empty

Every power series converges for some value of x. That is, the interval of convergence for a power series is never the empty set.

Although this fact has useful implications, it's actually pretty much a no-brainer. For example, take a look at the following power series:

$$\sum_{n=0}^{\infty} x^n = 1 + x + x^2 + x^3 + x^4 + \dots$$

When $x = 0$, this series evaluates to $1 + 0 + 0 + 0 + \dots$, so it obviously converges to 1. Similarly, take a peek at this power series:

$$\sum_{n=0}^{\infty} n(x+5)^n = (x+5) + 2(x+5)^2 + 3(x+5)^3 + 4(x+5)^4 + \dots$$

This time, when $x = -5$, the series converges to 0, just as trivially as the last example.

Note that in both of these examples, the series converges trivially at $x = a$ for a power series centered at a (see the beginning of "Power Series: Polynomials on Steroids").

Three varieties for the interval of convergence

Three possibilities exist for the interval of convergence of any power series:

✔ The series converges only when $x = a$.

✔ The series converges on some interval (open or closed at either end) centered at a.

✔ The series converges for all real values of x.

For example, suppose that you want to find the interval of convergence for:

$$\sum_{n=0}^{\infty} nx^n = x + 2x^2 + 3x^3 + 4x^4 + \ldots$$

This power series is centered at 0, so it converges when $x = 0$. Using the ratio test (see Chapter 12), you can find out whether it converges for any other values of x. To start out, set up the following limit:

$$\lim_{n \to \infty} \frac{(n+1)x^{n+1}}{nx^n}$$

To evaluate this limit, start out by x^n in the numerator and denominator:

$$= \lim_{n \to \infty} \frac{(n+1)x}{n}$$

Next, distribute to remove the parentheses in the numerator:

$$= \lim_{n \to \infty} \frac{nx + x}{n}$$

As it stands, this limit is of the form $\frac{\infty}{\infty}$, so apply L'Hospital's Rule (see Chapter 2), differentiating over the variable n:

$$\lim_{n \to \infty} x = x$$

From this result, the ratio test tells you that the series:

✔ Converges when $-1 < x < 1$

✔ Diverges when $x < -1$ and $x > 1$

✔ May converge or diverge when $x = 1$ and $x = -1$

Fortunately, it's easy to see what happens in these two remaining cases. Here's what the series looks like when $x = 1$:

$$\sum_{n=0}^{\infty} n(1)^n = 1 + 2 + 3 + 4 + \ldots$$

Clearly, the series diverges. Similarly, here's what it looks like when $x = -1$:

$$\sum_{n=0}^{\infty} n(-1)^{n} = -1 + 2 - 3 + 4 - \ldots$$

This alternating series swings wildly between negative and positive values, so it also diverges.

As a final example, suppose that you want to find the interval of convergence for the following series:

$$\sum_{n=0}^{\infty} (-1)^{n} \frac{x^{2n}}{(2n)!} = 1 - \frac{x^{2}}{2!} + \frac{x^{4}}{4!} - \frac{x^{6}}{6!} + \ldots$$

As in the last example, this series is centered at 0, so it converges when $x = 0$. The real question is whether it converges for other values of x. Because this is an alternating series, I apply the ratio test to the positive version of it to see whether I can show that it's absolutely convergent:

$$\lim_{n \to \infty} \frac{\dfrac{x^{2(n+1)}}{(2(n+1))!}}{\dfrac{x^{2n}}{(2n)!}}$$

First off, I want to simplify this a bit:

$$= \lim_{n \to \infty} \frac{\dfrac{x^{2n+2}}{(2n+2)!}}{\dfrac{x^{2n}}{(2n)!}}$$

$$= \lim_{n \to \infty} \frac{x^{2n+2}}{(2n+2)!} \cdot \frac{(2n)!}{x^{2n}}$$

Next, I expand out the exponents and factorials, as I show you in Chapter 12:

$$= \lim_{n \to \infty} \frac{x^{2n} x^{2}}{(2n+2)(2n+1)(2n)!} \cdot \frac{(2n)!}{x^{2n}}$$

At this point, a lot of canceling is possible:

$$= \lim_{n \to \infty} \frac{x^{2}}{(2n+2)(2n+1)} = 0$$

This time, the limit falls between -1 and 1 for all values of x. This result tells you that the series converges absolutely for all values of x, so the alternating series also converges for all values of x.

Expressing Functions as Series

In this section, you begin to explore how to express functions as infinite series. I begin by showing some examples of formulas that express sin x and cos x as series. These examples lead to a more general formula for expressing a wider variety of elementary functions as series.

This formula is the Maclaurin series, a simplified but powerful version of the more general Taylor series, which I introduce later in this chapter.

Expressing sin x as a series

Here's an odd formula that expresses the sine function as an alternating series:

$$\sin x = \sum_{x=0}^{\infty} (-1)^n \, \frac{x^{2n+1}}{(2n+1)!}$$

To make sense of this formula, use expanded notation:

$$\sin x = x - \frac{x^3}{3!} + \frac{x^5}{5!} - \frac{x^7}{7!} \cdots$$

Notice that this is a power series (which I discuss earlier in this chapter). To get a quick sense of how it works, here's how you can find the value of sin 0 by substituting 0 for x:

$$\sin 0 = 0 - \frac{0^3}{3!} + \frac{0^5}{5!} - \frac{0^7}{7!} + \ldots = 0$$

As you can see, the formula verifies what you already know: sin 0 = 0.

You can use this formula to approximate sin x for any value of x to as many decimal places as you like. For example, look what happens when you substitute 1 for x in the first four terms of the formula:

$$\sin 1 \approx 1 - \frac{1}{6} + \frac{1}{120} - \frac{1}{5{,}040}$$
$$\approx 0.841468$$

Note that the actual value of sin 1 to six decimal places is 0.841471, so this estimate is correct to five decimal places — not bad!

Table 13-1 shows the value of sin 3 approximated out to six terms. Note that the actual value of sin 3 is approximately 0.14112, so the six-term approximation is correct to three decimal places. Again, not bad, though not quite as good as the estimate for sin 1.

Table 13-1	Approximating the Value of sin 3	
# of Terms	Substitution	Approximation
1	3	3
2	$3 - \dfrac{3^3}{3!}$	−1.5
3	$3 - \dfrac{3^3}{3!} + \dfrac{3^5}{5!}$	0.525
4	$3 - \dfrac{3^3}{3!} + \dfrac{3^5}{5!} - \dfrac{3^7}{7!}$	0.09107
5	$3 - \dfrac{3^3}{3!} + \dfrac{3^5}{5!} - \dfrac{3^7}{7!} + \dfrac{3^9}{9!}$	0.14531
6	$3 - \dfrac{3^3}{3!} + \dfrac{3^5}{5!} - \dfrac{3^7}{7!} + \dfrac{3^9}{9!} - \dfrac{3^{11}}{11!}$	0.14087

As a final example, Table 13-2 shows the value of sin 10 approximated out to eight terms. The true value of sin 10 is approximately −0.54402, so by any standard this is a poor estimate. Nevertheless, if you continue to generate terms, this estimate continues to get better and better, to any level of precision you like. If you doubt this, notice that after five terms, the approximations are beginning to get closer to the actual value.

Table 13-2	Approximating the Value of sin 10	
# of Terms	Substitution	Approximation
1	10	10
2	$10 - \dfrac{10^3}{3!}$	−156.66667
3	$10 - \dfrac{10^3}{3!} + \dfrac{10^5}{5!}$	676.66667
4	$10 - \dfrac{10^3}{3!} + \dfrac{10^5}{5!} - \dfrac{10^7}{7!}$	−1307.460317
5	$10 - \dfrac{10^3}{3!} + \dfrac{10^5}{5!} - \dfrac{10^7}{7!} + \dfrac{10^9}{9!}$	1448.272
6	$10 - \dfrac{10^3}{3!} + \dfrac{10^5}{5!} - \dfrac{10^7}{7!} + \dfrac{10^9}{9!} - \dfrac{10^{11}}{11!}$	−1056.938
7	$10 - \dfrac{10^3}{3!} + \dfrac{10^5}{5!} - \dfrac{10^7}{7!} + \dfrac{10^9}{9!} - \dfrac{10^{11}}{11!} + \dfrac{10^{13}}{13!}$	548.966
8	$10 - \dfrac{10^3}{3!} + \dfrac{10^5}{5!} - \dfrac{10^7}{7!} + \dfrac{10^9}{9!} - \dfrac{10^{11}}{11!} + \dfrac{10^{13}}{13!} - \dfrac{10^{15}}{15!}$	−215.750

Expressing cos x as a series

In the previous section, I show you a formula that expresses the value of sin x for all values of x as an infinite series. Differentiating both sides of this formula leads to a similar formula for cos x:

$$\frac{d}{dx}\sin x = \frac{d}{dx}x - \frac{d}{dx}\frac{x^3}{3!} + \frac{d}{dx}\frac{x^5}{5!} - \frac{d}{dx}\frac{x^7}{7!} + \cdots$$

Now, evaluate these derivatives:

$$\cos x = 1 - 3\frac{x^2}{3!} + 5\frac{x^4}{5!} - 7\frac{x^6}{7!} + \cdots$$

Finally, simplify the result a bit:

$$\cos x = 1 - \frac{x^2}{2!} + \frac{x^4}{4!} - \frac{x^6}{6!} + \cdots$$

As you can see, the result is another power series (which I discuss earlier in this chapter). Here's how you write it by using sigma notation:

$$\cos x = \sum_{n=0}^{\infty} (-1)^n \frac{x^{2n}}{(2n)!}$$

To gain some confidence that this series really works as advertised, note that the substitution $x = 0$ provides the correct equation cos 0 = 1. Furthermore, substituting $x = 1$ into the first four terms gives you the following approximation:

$$\cos 1 \approx 1 - \frac{1}{2} + \frac{1}{24} - \frac{1}{720} = 0.540\overline{277}$$

This estimate is accurate to four decimal places.

Introducing the Maclaurin Series

In the last two sections, I show you formulas for expressing both sin x and cos x as infinite series. You may begin to suspect that there's some sort of method behind these formulas. Without further ado, here it is:

$$f(x) = \sum_{n=0}^{\infty} \frac{f^{(n)}(0)}{n!} x^n$$

Behold the *Maclaurin series*, a simplified version of the much-heralded Taylor series, which I introduce in the next section.

The notation $f^{(n)}$ means "the nth derivative of f." This should become clearer in the expanded version of the Maclaurin series:

$$f(x) = f(0) + f'(0)x + \frac{f''(0)}{2!}x^2 + \frac{f'''(0)}{3!}x^3 + \cdots$$

The Maclaurin series is the template for the two formulas I introduce earlier in this chapter. It allows you to express many other functions as power series by following these steps:

1. **Find the first few derivatives of the function until you recognize a pattern.**

2. **Substitute 0 for x into each of these derivatives.**

3. **Plug these values, term by term, into the formula for the Maclaurin series.**

4. **If possible, express the series in sigma notation.**

For example, suppose that you want to find the Maclaurin series for e^x.

1. **Find the first few derivatives of e^x until you recognize a pattern:**

$$f'(x) = e^x$$
$$f''(x) = e^x$$
$$f'''(x) = e^x$$
$$\dots$$
$$f^{(n)}(x) = e^x$$

2. **Substitute 0 for x into each of these derivatives.**

$$f'(0) = e^0$$
$$f''(0) = e^0$$
$$f'''(0) = e^0$$
$$\dots$$
$$f^{(n)}(x) = e^0$$

3. **Plug these values, term by term, into the formula for the Maclaurin series:**

$$e^x = e^0 + e^0 x + \frac{e^0}{2!} x^2 + \frac{e^0}{3!} x^3 + \dots$$
$$= 1 + x + \frac{x^2}{2} + \frac{x^3}{6} + \dots$$

4. **If possible, express the series in sigma notation:**

$$e^x = \sum_{n=0}^{\infty} \frac{x^n}{n!}$$

To check this formula, use it to estimate e^0 and e^1 by substituting 0 and 1, respectively, into the first six terms:

$$e^0 = 1 + 0 + 0 + 0 + 0 + 0 + \dots = 1$$
$$e^1 \approx 1 + 1 + \frac{1}{2} + \frac{1}{6} + \frac{1}{24} + \frac{1}{120} = 2.7166(\text{repeating})$$

This exercise nails e^0 exactly, and approximates e^1 to two decimal places. And, as with the formulas for $\sin x$ and $\cos x$ that I show you earlier in this chapter, the Maclaurin series for e^x allows you to calculate this function for any value of x to any number of decimal places.

As with the other formulas, however, the Maclaurin series for e^x works best when x is close to 0. As x moves away from 0, you need to calculate more terms to get the same level of precision.

But now, you can begin to see why the Maclaurin series tends to provide better approximations for values close to 0: The number 0 is "hardwired" into the formula as $f(0)$, $f'(0)$, $f''(0)x$, and so forth.

Figure 13-1 illustrates this point. The first graph shows $\sin x$ approximated by using the first two terms of the Maclaurin series — that is, as the third-degree polynomial $x - \frac{x^3}{3!}$. The subsequent graph shows an approximation of $\sin x$ with four terms.

A tale of three series

It's easy to get confused about the three categories of series that I discuss in this chapter. Here's a helpful way to think about them:

 ✔ The *power series* is a subcategory of infinite series.

 ✔ The *Taylor series* (named for mathematician Brook Taylor) is a subcategory of power series.

 ✔ The *Maclaurin series* (named for mathematician Colin Maclaurin) is a subcategory of Taylor series.

After you have that down, consider that the power series has two basic forms:

 ✔ The *specific form*, which is centered at zero, so *a* drops out of the expression.

 ✔ The *general form*, which isn't centered at zero, so *a* is part of the expression.

Furthermore, each of the other two series uses one of these two forms of the power series:

 ✔ The Maclaurin series uses the specific form, so it's:

  • Less powerful

  • Simpler to work with

 ✔ The Taylor series uses the general form, so it's:

  • More powerful

  • Harder to work with

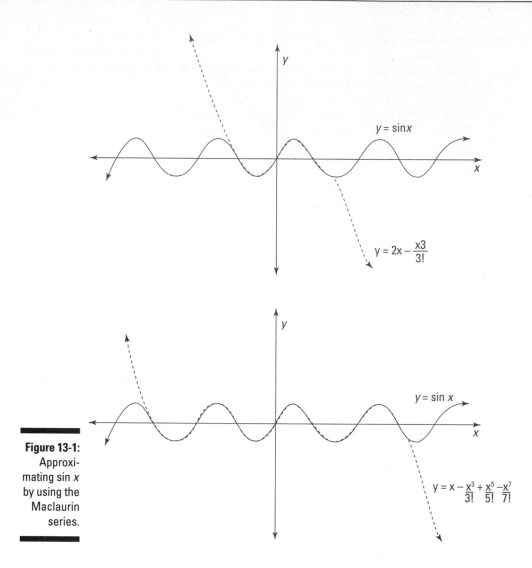

Figure 13-1:
Approxi-
mating sin x
by using the
Maclaurin
series.

As you can see, each successive approximation improves upon the previous one. Furthermore, each equation tends to provide its best approximation when x is close to 0.

Introducing the Taylor Series

Like the Maclaurin series (which I introduce in the previous section), the Taylor series provides a template for representing a wide variety of functions as power series.

In fact, the Taylor series is really a more general version of the Maclaurin series. The advantage of the Maclaurin series is that it's a bit simpler to work with. The advantage to the Taylor series is that you can tailor it to obtain a better approximation of many functions.

Without further ado, here's the Taylor series in all its glory:

$$f(x) = \sum_{n=0}^{\infty} \frac{f^{(n)}(a)}{n!} (x-a)^n$$

As with the Maclaurin series, the Taylor series uses the notation $f^{(n)}$ to indicate the nth derivative. Here's the expanded version of the Taylor series:

$$f(x) = f(a) + f'(a)(x-a) + \frac{f''(a)}{2!}(x-a)^2 + \frac{f'''(a)}{3!}(x-a)^3 + \dots$$

Notice that the Taylor series includes the variable a, which isn't found in the Maclaurin series. Or, more precisely, in the Maclaurin series, $a = 0$, so it drops out of the expression.

The explanation for this variable can be found earlier in this chapter, in "Power Series: Polynomials on Steroids." In that section, I show you two forms of the power series:

 ✔ A simpler form centered at 0, which corresponds to the Maclaurin series

 ✔ A more general form centered at a, which corresponds to the Taylor series

In the next section, I show you the advantages of working with this extra variable.

Computing with the Taylor series

The presence of the variable a makes the Taylor series more complex to work with than the Maclaurin series. But this variable provides the Taylor series with greater flexibility, as the next example illustrates.

In "Expressing Functions as Series" earlier in this chapter, I attempt to approximate the value of sin 10 with the Maclaurin series. Unfortunately, taking this calculation out to eight terms still results in a poor estimate. This problem occurs because the Maclaurin series always takes a default value of $a = 0$, and 0 isn't close enough to 10.

This time, I use only four terms of the Taylor series to make a much better approximation. The key to this approximation is a shrewd choice for the variable a:

Let $a = 3\pi$

This choice has two advantages: First, this value of a is close to 10 (the value of x), which makes for a better approximation. Second, it's an easy value for calculating sines and cosines, so the computation shouldn't be too difficult.

To start off, substitute 10 for x and 3π for a in the first four terms of the Taylor series:

$$\sin 10 = \sin 3\pi + (\sin' 3\pi)(10 - 3\pi) + \frac{(\sin'' 3\pi)(10 - 3\pi)^2}{2!} + \frac{(\sin''' 3\pi)(10 - 3\pi)^3}{3!}$$

Next, substitute in the first, second, and third derivatives of the sine function and simplify:

$$= \sin 3\pi + (\cos 3\pi)(0.5752) - \frac{(\sin 3\pi)(0.5752)^2}{2!} - \frac{(\cos 3\pi)(0.5752)^3}{3!}$$

The good news is that $\sin 3\pi = 0$, so the first and third terms fall out:

$$= (\cos 3\pi)(0.5752) - \frac{(\cos 3\pi)(0.5752)^3}{3!}$$

At this point, you probably want to grab your calculator:

$$= -1\,(0.5752) - -\frac{1}{6}\,(0.5752)^3$$

$$= -0.5752 + 0.0317 = -0.5434$$

This approximation is correct to two decimal places — quite an improvement over the estimate from the Maclaurin series!

Examining convergent and divergent Taylor series

Earlier in this chapter, I show you how to find the interval of convergence for a power series — that is, the set of x values for which that series converges.

Because the Taylor series is a form of power series, you shouldn't be surprised that every Taylor series also has an interval of convergence. When this interval is the entire set of real numbers, you can use the series to find the value of $f(x)$ for every real value of x.

However, when the interval of convergence for a Taylor series is bounded — that is, when it diverges for some values of x — you can use it to find the value of $f(x)$ *only* on its interval of convergence.

For example, here are the three important Taylor series that I've introduced so far in this chapter:

$$\sin x = \sum_{n=0}^{\infty} (-1)^n \frac{x^{2n+1}}{(2n+1)!} = x - \frac{x^3}{3!} + \frac{x^5}{5!} - \frac{x^7}{7!} + \ldots$$

$$\cos x = \sum_{n=0}^{\infty} (-1)^n \frac{x^{2n}}{(2n)!} = 1 - \frac{x^2}{2!} + \frac{x^4}{4!} - \frac{x^6}{6!} + \ldots$$

$$e^x = \sum_{n=0}^{\infty} \frac{x^n}{n!} = 1 + x + \frac{x^2}{2!} + \frac{x^3}{3!} + \ldots$$

All three of these series converge for all real values of x (you can check this by using the ratio test, as I show you earlier in this chapter), so each equals the value of its respective function.

Now, consider the following function:

$$f(x) = \frac{1}{1-x}$$

I express this function as a Maclaurin series, using the steps that I outline earlier in this chapter in "Expressing Functions as Series":

1. **Find the first few derivatives of $f(x) = \frac{1}{1-x}$ until you recognize a pattern:**

$$f'(x) = \frac{1}{1-x^2}$$

$$f''(x) = \frac{2}{(1-x)^3}$$

$$f'''(x) = \frac{6}{(1-x)^4}$$

...

$$f^{(n)}(x) = \frac{n!}{(1-x)^{n+1}}$$

2. **Substitute 0 for x into each of these derivatives:**

$f'(0) = 1$

$f''(0) = 2$

$f'''(0) = 6$

...

$f^{(n)}(0) = n!$

3. **Plug these values, term by term, into the formula for the Maclaurin series:**

$$\frac{1}{1-x} = f(0) + f'(0)x + \frac{f''(0)}{2!}x^2 + \frac{f'''(0)}{3!}x^3 + \dots$$

$$= 1 + x + x^2 + x^3 + \dots$$

4. **If possible, express the series in sigma notation:**

$$\frac{1}{1-x} = \sum_{n=0}^{\infty} x^n = 1 + x + x^2 + x^3$$

To test this formula, I use it to find $f(x)$ when $x = \frac{1}{2}$.

$$f\left(\frac{1}{2}\right) = 1 + \frac{1}{2} + \frac{1}{4} + \frac{1}{8} + \dots = 2$$

You can test the accuracy of this expression by substituting $\frac{1}{2}$ into $\frac{1}{1-x}$:

$$f\left(\frac{1}{2}\right) = \frac{1}{1 - \frac{1}{2}} = 2$$

As you can see, the formula produces the correct answer. Now, I try to use it to find $f(x)$ when $x = 5$, noting that the correct answer should be $\frac{1}{1-5} = -\frac{1}{4}$:

$$f(5) = 1 + 5 + 25 + 125 + \dots = \infty \quad \text{WRONG!}$$

What happened? This series converges only on the interval $(-1, 1)$, so the formula produces only the value $f(x)$ when x is in this interval. When x is outside this interval, the series diverges, so the formula is invalid.

Expressing functions versus approximating functions

It's important to be crystal clear in your understanding about the difference between two key mathematical practices:

- ✔ *Expressing* a function as an infinite series
- ✔ *Approximating* a function by using a finite number of terms of series

Both the Taylor series and the Maclaurin series are variations of the power series. You can think of a power series as a polynomial with infinitely many terms. Also, recall that the Maclaurin series is a specific form of the more general Taylor series, arising when the value of a is set to 0.

Every Taylor series (and, therefore, every Maclaurin series) provides the exact value of a function for all values of x where that series converges. That is, for any value of x on its interval of convergence, a Taylor series converges to $f(x)$.

In practice, however, adding up an infinite number of terms simply isn't possible. Nevertheless, you can approximate the value of $f(x)$ by adding up a finite number from the appropriate Taylor series. You do this earlier in the chapter to estimate the value of sin 10 and other expressions.

An expression built from a finite number of terms of a Taylor series is called a *Taylor polynomial, $T_n(x)$*. Like other polynomials, a Taylor polynomial is identified by its degree. For example, here's the fifth-degree Taylor polynomial, $T_5(x)$, that approximates e^x:

$$e^x \approx 1 + x + \frac{x^2}{2!} + \frac{x^3}{3!} + \frac{x^4}{4!} + \frac{x^5}{5!}$$

Generally speaking, a higher-degree polynomial results in a better approximation. And because this polynomial comes from the Maclaurin series, where $a = 0$, it provides a much better estimate for values of e^x when x is near 0. For the value of e^x when x is near 100, however, you get a better estimate by using a Taylor polynomial for e^x with $a = 100$:

$$e^x \approx e^{100} + e^{100}(x - 100) + \frac{e^{100}}{2!}(x - 100)^2 + \frac{e^{100}}{3!}(x - 100)^3 + \frac{e^{100}}{4!}(x - 100)^4 +$$
$$\frac{e^{100}}{5!}(x - 100)^5$$

To sum up, remember the following:

✔ A convergent Taylor series expresses the exact value of a function.

✔ A Taylor polynomial, $T_n(x)$, from a convergent series approximates the value of a function.

Calculating error bounds for Taylor polynomials

In the previous section, I discuss how a Taylor polynomial approximates the value of a function:

$$f(x) \approx T_n(x)$$

In many cases, it's helpful to measure the accuracy of an approximation. This information is provided by the *Taylor remainder term:*

$$f(x) = T_n(x) + R_n(x)$$

Notice that the addition of the remainder term $R_n(x)$ turns the approximation into an equation. Here's the formula for the remainder term:

$$R_n(x) = \frac{f^{(n+1)}(c)}{(n+1)!}(x-a)^{n+1} \qquad c \text{ between } a \text{ and } x$$

It's important to be clear that this equation is true for one *specific* value of c on the interval between a and x. It does *not* work for just any value of c on that interval.

Ideally, the remainder term gives you the precise difference between the value of a function and the approximation $T_n(x)$. However, because the value of c is uncertain, in practice the remainder term really provides a worst-case scenario for your approximation.

An example should help to make this idea clear. I use the sixth-degree Taylor polynomial for $\cos x$:

$$\cos x \approx T_6(x) = 1 - \frac{x^2}{2!} + \frac{x^4}{4!} - \frac{x^6}{6!}$$

Suppose that I use this polynomial to approximate $\cos 1$:

$$\cos 1 \approx T_6(1) = 1 - \frac{1}{2} + \frac{1}{24} - \frac{1}{720}$$
$$= 0.540\overline{277}$$

How accurate is this approximation likely to be? To find out, utilize the remainder term:

$$\cos 1 = T_6(x) + R_6(x)$$

Adding the associated remainder term changes this approximation into an equation. Here's the formula for the remainder term:

$$R_6(x) = \frac{\cos^{(7)} c}{7!}(x-0)^7$$
$$= \frac{\sin c}{5,040} x^7 \qquad c \text{ between } 0 \text{ and } x$$

So, substituting 1 for x gives you:

$$R_6(1) = \frac{\sin c}{5,040} \qquad c \text{ between } 0 \text{ and } 1$$

At this point, you're apparently stuck, because you don't know the value of $\sin c$. However, the sin function always produces a number between -1 and 1, so you can narrow down the remainder term as follows:

$$-\frac{1}{5,040} \leq R_6(1) \leq \frac{1}{5,040}$$

Note that $\frac{1}{5,040} \approx 0.0001984$, so the approximation of cos 1 given by the $T_6(1)$ is accurate to within 0.0001984 in either direction. And, in fact, cos 1 $\approx$ 0.540302, so:

$$\cos 1 - T_6(1) \approx 0.540302 - 0.540278 = 0.000024$$

As you can see, the approximation is within the error bounds predicted by the remainder term.

Understanding Why the Taylor Series Works

The best way to see why the Taylor series works is to see how it's constructed in the first place. If you read through this chapter until this point, you should be ready to go.

To make sure that you understand every step along the way, however, I construct the Maclaurin series, which is just a tad more straightforward. This construction begins with the key assumption that a function can be expressed as a power series in the first place:

$$f(x) = c_0 + c_1 x + c_2 x^2 + c_3 x^3 + \dots$$

The goal now is to express the coefficients on the right side of this equation in terms of the function itself. To do this, I make another relatively safe assumption that 0 is in the domain of $f(x)$. So when $x = 0$, all but the first term of the series equal 0, leaving the following equation:

$$f(0) = c_0$$

This process gives you the value of the coefficient c_0 in terms of the function. Now, differentiate $f(x)$:

$$f'(x) = c_1 + 2c_2 x + 3c_3 x^2 + 4c_4 x^3 \dots$$

At this point, when $x = 0$, all the x terms drop out:

$$f'(0) = c_1$$

So you have another coefficient, c_1, expressed in terms of the function. To continue, differentiate $f'(x)$:

$$f''(x) = 2c_2 + 6c_3 x + 12c_4 x^2 + 20c_5 x^3 + \dots$$

Again, when $x = 0$, the x terms disappear:

$$f''(0) = 2c_2$$

$$\frac{f''(0)}{2} = c_2$$

By now, you're probably noticing a pattern: You can always get the value of the next coefficient by differentiating the previous equation and substituting 0 for x into the result:

$$f'''(x) = 6c_3 + 24c_4 x + 60c_5 x^2 + 120c_6 x^3 + \ldots$$

$$f'''(0) = 6c_3$$

$$\frac{f'''(0)}{6} = c_3$$

Furthermore, the coefficients also have a pattern:

$$c_0 = f(0)$$

$$c_1 = f'(0)$$

$$c_2 = \frac{f''(0)}{2!}$$

$$c_3 = \frac{f'''(0)}{3!}$$

$$\ldots$$

$$c_n = \frac{f^{(n)}(0)}{n!}$$

Substituting these coefficients into the original equation results in the familiar Maclaurin series from earlier in this chapter:

$$f(x) = f(0) + f'(0)x + \frac{f'(0)}{2!} x^2 + \frac{f'''(0)}{3!} x^3 + \ldots$$

To construct the Taylor series, use a similar line of reasoning, starting with the more general form of the power series:

$$f(x) = c_0 + c_1(x - a) + c_2(x - a)^2 + c_3(x - a)^3 + \ldots$$

In this case, setting $x = a$ gives you the first coefficient:

$$f(a) = c_0$$

Continue to find coefficients by differentiating $f(x)$ and then repeating the process.

Part V
Advanced Topics

The 5th Wave   By Rich Tennant

Sorry. I just find rotating my head helps me to relax during the test.

In this part . . .

You get a glimpse of what lies beyond Calculus II. I give you an overview of the next two semesters of math: Calculus III (the study of calculus in three or more dimensions) and Differential Equations (equations with derivatives mixed in as variables).

Chapter 14

Multivariable Calculus

Space, as Captain Kirk says during the opening credits of the television series *Star Trek,* is the final frontier. Multivariable calculus (also known as Calculus III) focuses on techniques for doing calculus in space — that is, in three dimensions.

Mathematicians have a variety of terms for three dimensions: 3-D, 3-space, and R^3 are the most common. Whatever you call it, adding a dimension makes multivariable calculus more interesting and useful, but also a bit more tricky than single variable calculus.

In this chapter, I give you a quick introduction to multivariable calculus, touching on the highlights usually taught in a Calculus III class. First, I show you how vectors provide a method for linking a value with a direction. Next, I introduce you to three different 3-D coordinate systems: 3-D Cartesian coordinates, cylindrical coordinates, and spherical coordinates.

Building on this understanding of 3-D, I discuss functions of more than one variable, focusing on the function of two variables $z = f(x, y)$. With an understanding of multivariable functions, I proceed to introduce you to the two most important concepts in multivariable calculus: partial derivatives and multiple integrals.

By the end of this chapter, you'll have a great platform from which to begin Calculus III.

Visualizing Vectors

Vectors are used to link a real number (called a *scalar*) with a direction on the plane or in space. They're useful for navigation, where knowing what direction you're sailing or flying in is important. Vectors also get a lot of play in physics, where forces that push and pull are also directional. And, as you may have guessed, Calculus III is chock full of vectors.

In this section, I introduce you to this important concept. Although I keep this discussion in two dimensions, vectors are commonly used in three dimensions as well.

Understanding vector basics

A simple way to think of a vector is as an arrow that has both length and direction. By convention, a vector starts at the origin of the Cartesian plane (0, 0) and extends a certain length in some direction. Figure 14-1 shows a variety of vectors.

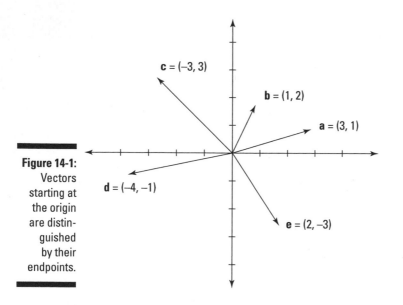

Figure 14-1:
Vectors starting at the origin are distinguished by their endpoints.

As you can see, when a vector begins at the origin, its two *components* (its *x* and *y* values) correspond to the Cartesian coordinates of its endpoint. For example, the vector that begins at (0, 0) and ends at (3, 1) is distinguished as the vector <3, 1>.

Don't confuse a vector <*x, y*> with its corresponding Cartesian pair (*x, y*), which is just a point.

By convention, vectors are labeled in books with boldfaced lowercase letters: **a**, **b**, **c**, and so forth (see Figure 14-1). But when you're working with vectors on paper, most teachers are happy to see you replace the boldface with a little line or arrow over the letter.

Displacing a vector from the origin doesn't change its value. For example, Figure 14-2 shows the vector <2, 3> with a variety of starting points.

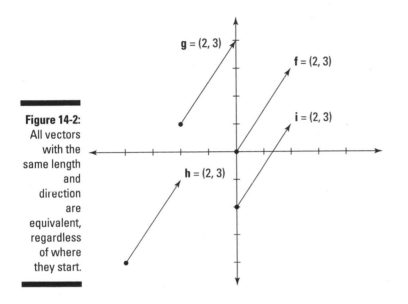

Figure 14-2:
All vectors with the same length and direction are equivalent, regardless of where they start.

Calculate the coordinates of a vector starting at (x_1, y_1) and ending at (x_2, y_2) as <$x_2 - x_1, y_2 - y_1$>. For example, here's how to calculate the coordinates of vectors **g**, **h**, and **i** in Figure 14-2:

$$\mathbf{g} = <0 - -2, 4 - 1> = <2, 3>$$
$$\mathbf{h} = <-2 - -4, -1 - -4> = <2, 3>$$
$$\mathbf{i} = <2 - 0, 1 - -2> = <2, 3>$$

As you can see, regardless of the starting and ending points, these vectors are all equivalent to each other and to **f**.

Distinguishing vectors and scalars

Just as Eskimos have tons of words for snow and Italians have even more words for pasta, mathematicians have bunches of words for real numbers. When you began algebra, you found out quickly that a *constant* or a *coefficient* was just a real number in a specific context.

Similarly, when you're discussing vectors and want to refer to a real number, use the word *scalar*. Only the name is different — deep in its heart, a scalar knows only too well that it's just a good old-fashioned real number.

Some types of vector calculations produce new vectors, while others result in scalars. As I introduce these calculations throughout the next section, I tell you whether to expect a vector or a scalar as a result.

Calculating with vectors

Vectors are commonly used to model forces such as wind, sea current, gravity, and electromagnetism. Vector calculations are essential for all sorts of problems where forces collide. In this section, I give you a taste of how some simple calculations with vectors are accomplished.

Calculating magnitude

The length of a vector is called its *magnitude*. The notation for absolute value (| |) is also used for the magnitude of a vector. For example, |**v**| refers to magnitude of the vector **v**. (By the way, some textbooks represent magnitude with double bars [| | | |] instead of single bars. Either way, the meaning is the same.)

What's in a name?

A scalar is just a fancy word for a real number. The name arises because a scalar *scales* a vector — that is, it changes the scale of a vector. For example, the real number 2 scales the vector **v** by a factor of 2, so that 2**v** is twice as long as **v**. You find out more about scalar multiplication in "Calculating with vectors."

Calculate the magnitude of a vector **v** = <x, y> by using a variation of the distance formula. This formula is itself a variation of the trusty Pythagorean theorem:

$$|\mathbf{v}| = \sqrt{x^2 + y^2}$$

For example, calculate the magnitude of the vector **n** = <4, –3> as follows:

$$|\mathbf{n}| = \sqrt{4^2 + (-3)^2} = 5$$

As you can see, the use of the absolute value bars for the magnitude of vectors is appropriate: Magnitude, like all other distances, is always measured as a nonnegative value. The magnitude of a vector is the distance from the origin of a graph to its tip, just as the absolute value of a number is the distance from 0 on a number line to that number.

The magnitude of a vector is a *scalar*.

Scalar multiplication

Multiplying a vector by a scalar is called *scalar multiplication*. To perform scalar multiplication, multiply the scalar by each component of the vector. Here's how you multiply the vector **v** = <x, y> by the scalar k:

$$k\mathbf{v} = k<x, y> = <kx, ky>$$

For example, here's how you multiply the vector **p** = <3, 5> by the scalars 2, –4, and $\frac{1}{3}$:

$$2\mathbf{p} = 2<3, 5> = <6, 10>$$

$$-4\mathbf{p} = -4<3, 5> = <-12, -20>$$

$$\tfrac{1}{3}\mathbf{p} = \tfrac{1}{3}<3, 5> = <1, \tfrac{5}{3}>$$

When you multiply a vector by a scalar, the result is a *vector*.

Geometrically speaking, scalar multiplication achieves the following.

- ✔ Scalar multiplication by a positive number other than 1 changes the magnitude of the vector but not its direction.
- ✔ Scalar multiplication by –1 reverses its direction but doesn't change its magnitude.
- ✔ Scalar multiplication by any other negative number both reverses the direction of the vector and changes its magnitude.

Scalar multiplication can change the magnitude of a vector by either increasing it or decreasing it.

- ✔ Scalar multiplication by a number greater than 1 or less than –1 *increases* the magnitude of the vector.
- ✔ Scalar multiplication by a fraction between –1 and 1 *decreases* the magnitude of the vector.

For example, the vector 2**p** is twice as long as **p**, the vector $\frac{1}{2}$ **p** is half as long as **p**, and the vector –**p** is the same length as **p** but extends in the opposite direction from the origin (as shown in Figure 14-3).

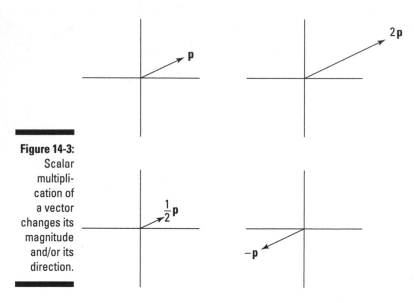

Figure 14-3: Scalar multiplication of a vector changes its magnitude and/or its direction.

Finding the unit vector

Every vector has a corresponding *unit vector,* which has the same direction as that vector but a magnitude of 1. To find the unit vector **u** of the vector **v** = <*x, y*>, divide that vector by its magnitude as follows:

$$\mathbf{u} = \frac{\mathbf{v}}{|\mathbf{v}|}$$

Note that this formula uses scalar multiplication, as I show you in the preceding section, because the numerator is a vector and the denominator is a scalar.

As you may guess from its name, the unit vector is a *vector.*

For example, to find the unit vector **u** of the vector **q** = <−2, 1>, first calculate its magnitude | **q** | as I show you earlier in this section:

$$|\mathbf{q}| = \sqrt{(-2)^2 + 1^2} = \sqrt{5}$$

Now, use the previous formula to calculate the unit vector:

$$\mathbf{u} = \frac{<-2,1>}{\sqrt{5}} = <-\frac{2}{\sqrt{5}}, \frac{1}{\sqrt{5}}>$$

You can check that the magnitude of resulting vector **u** really is 1 as follows:

$$|\mathbf{u}| = \sqrt{\left(-\frac{2}{\sqrt{5}}\right)^2 + \left(\frac{1}{\sqrt{5}}\right)^2}$$

$$= \sqrt{\frac{4}{5} + \frac{1}{5}} = 1$$

Adding and subtracting vectors

Add and subtract vectors component by component, as follows:

$$<x_1, y_1> + <x_2, y_2> = <x_1 + x_2, y_1 + y_2>$$

$$<x_1, y_1> - <x_2, y_2> = <x_1 - x_2, y_1 - y_2>$$

For example, if **r** = <−1, 3> and **s** = <4, 2>, here's how to add and subtract these vectors:

$$\mathbf{r} + \mathbf{s} = <-1, 3> + <4, 2> = <-1 + 4, 3 + 2> = <3, 5>$$

$$\mathbf{r} - \mathbf{s} = <-1, 3> - <4, 2> = <-1 - 4, 3 - 2> = <-5, 1>$$

When you add or subtract two vectors, the result is a *vector.*

Geometrically speaking, the net effects of vector addition and subtraction are shown in Figure 14-4. In this example, the endpoint of **r** + **s** is equivalent to the endpoint of **s** when **s** begins at the endpoint of **r**. Similarly, the endpoint of **r** − **s** is equivalent to the endpoint of −**s** — that is, <−4, −2> — when −**s** begins at the endpoint of **r**.

Figure 14-4:
Add and subtract vectors on the graph by beginning one vector at the endpoint of another vector.

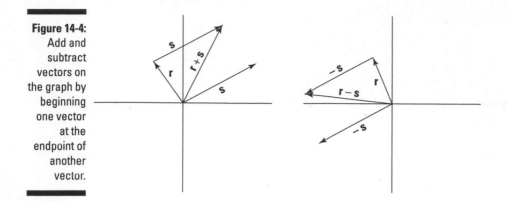

Leaping to Another Dimension

Multivariable calculus is all about three (or more) dimensions. In this section, I show you three important systems for plotting points in 3-D.

Understanding 3-D Cartesian coordinates

The three-dimensional (3-D) Cartesian coordinate system (also called 3-D rectangular coordinates) is the natural extension of the 2-D Cartesian graph. The key difference is the addition of a third axis, the z-axis, extending perpendicularly through the origin.

Drawing a 3-D graph in two dimensions is kind of tricky. To get a better sense about how to think in 3-D, hold up Figure 14-5 where you can compare it with the interior corner of a room (not a round room!). Note the following:

   ✔ The x-axis corresponds to where the left-hand wall meets the floor.

   ✔ The y-axis corresponds to where the right-hand wall meets the floor.

   ✔ The z-axis corresponds to where the two walls meet.

Just as the 2-D Cartesian graph is divided into four quadrants, the 3-D graph is divided into eight *octants*. From your perspective as you look at the graph, you're standing inside the first octant, where all values of x, y, and z are positive.

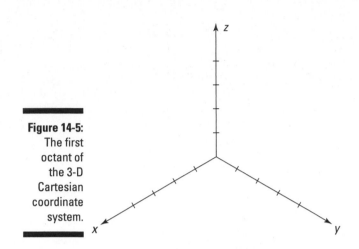

Figure 14-6 shows the complete 3-D Cartesian system with the point (1, 2, 5) plotted. In similarity with regular Cartesian coordinates, you plot this point by counting 1 unit in the positive x direction, and then 2 units in the positive y direction, and finally 5 units in the positive z direction.

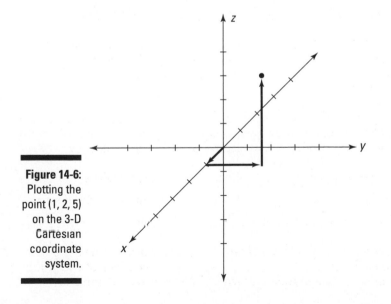

Using alternative 3-D coordinate systems

In Chapter 2, I discuss polar coordinates, an alternative to the Cartesian graph. Polar coordinates are useful because they allow you to express and solve a variety of problems more easily than Cartesian coordinates.

In this section, I show you two alternatives to the 3-D Cartesian coordinate system: cylindrical coordinates and spherical coordinates. As with polar coordinates, both of these systems give you greater flexibility to solve a wider range of problems.

Cylindrical coordinates

You probably remember polar coordinates from Pre-Calculus or maybe even Calculus I. Like the Cartesian coordinate system, the polar coordinate system assigns a pairing of values to every point on the plane. Unlike the Cartesian coordinate system, however, these values aren't dependent upon two perpendicular axes (though these axes are often drawn in to make the graph more readable). The key axis is the *horizontal axis,* which corresponds to the positive *x*-axis in Cartesian coordinates.

While a Cartesian pair is of the form (x, y), polar coordinates use (r, θ). *Cylindrical coordinates* are simply polar coordinates with the addition of a vertical *z*-axis extending from the origin, as in 3-D Cartesian coordinates (see "Understanding 3-D Cartesian coordinates" earlier in this chapter). Every point in space is assigned a set of cylindrical coordinates of the form (r, θ, z).

Here's what you need to know:

- ✔ The variable r measures the distance from the z-axis to that point.

- ✔ The variable θ measures angular distance from the horizontal axis. This angle is measured in radians rather than degrees, so that $2\pi = 360°$. (See Chapter 2 for more about radians.)

- ✔ The variable z measures the distance from that point to the xy-plane.

When plotting cylindrical coordinates, plot the first coordinates (r and θ) just as you would for polar coordinates (see Chapter 2). Then plot the z-coordinate as you would for 3-D Cartesian coordinates.

Figure 14-7 shows you how to plot the point $(3, \frac{\pi}{2}, 2)$ in cylindrical coordinates:

1. **Count 3 units to the right of the origin on the horizontal axis (as you would when plotting polar coordinates).**

2. **Travel counterclockwise along the arc of a circle until you reach the line drawn at a $\frac{\pi}{2}$-angle from the horizontal axis (again, as with polar coordinates).**

3. **Count 2 units above the plane and plot your point there.**

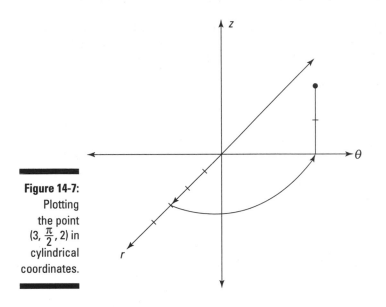

Spherical coordinates

Spherical coordinates are used — with slight variation — to measure latitude, longitude, and altitude on the most important sphere of them all, the planet Earth.

Every point in space is assigned a set of spherical coordinates of the form (ρ, θ, ϕ). In case you're not in a sorority or fraternity, ρ is the lowercase Greek letter *rho*, θ is the lowercase Greek letter *theta* (commonly used in math to represent an angle), ϕ is the lower-case Greek letter *phi*, which is commonly pronounced either "fee" or "fye" (but never "toe" or "fum").

The coordinate ρ corresponds to altitude. On the Earth, altitude is measured as the distance above or below sea level. In spherical coordinates, however, altitude indicates how far in space a point is from the origin.

The coordinate θ corresponds to longitude: A measurement of angular distance from the horizontal axis.

The coordinate φ corresponds to latitude. On the Earth, latitude is measured as angular distance from the equator. In spherical coordinates, however, latitude is measured as the angular distance from the north pole.

Plotting φ can be tricky at first. To get a feel for it, picture a globe and imagine traveling up and down along a single longitude line. Notice that as you travel, your latitude keeps changing, so

- At the north pole, φ = 0
- At the equator, φ = $\frac{\pi}{2}$
- At the south pole, φ = π

Some textbooks substitute the Greek letter ρ *(rho)* for *r*. Either way, the coordinate means the same thing: altitude, which is the distance of a point from the origin. In other textbooks, the order of the last two coordinates is changed around. Make sure that you know which convention your book uses.

Figure 14-8 shows you how to plot a point in spherical coordinates. For example, suppose that you want to plot the point (4, $\frac{\pi}{2}$, $\frac{3\pi}{4}$). Follow these steps to do that:

1. **Count 4 units to the right of the origin on the horizontal axis.**

2. **Travel counterclockwise along the arc of a circle until you reach the line drawn at a $\frac{\pi}{2}$-angle from the horizontal axis (again, as with polar coordinates).**

3. **Imagine a single longitude line arcing from the north pole of a sphere through the point on the equator where you are right now and onward to the south pole.**

4. **Travel down to the line of latitude at an angular distance of $\frac{3\pi}{4}$ from the north pole — that is, halfway between the equator and the south pole — and plot your point there.**

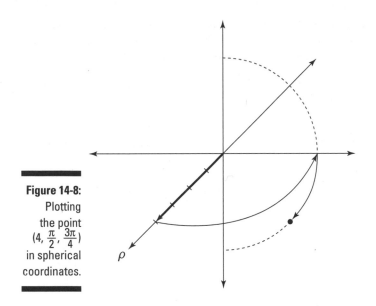

Figure 14-8:
Plotting
the point
$(4, \frac{\pi}{2}, \frac{3\pi}{4})$
in spherical
coordinates.

Functions of Several Variables

You know from algebra that a function $y = f(x)$ is basically a mathematical machine for turning one number into another. The variable x is the input variable and y is the output variable so that every value of x gives you no more than one y value.

When you graph a curve, you can use the vertical-line test to make sure that it's a function: Any vertical line that intersects a function intersects it at exactly one point, as illustrated in Figure 14-9.

These concepts related to functions also carry over into functions of more than one variable. For example, here are some functions of two variables:

$$z = 2x + y + 5$$
$$z = \sin x + y + \sqrt{y}$$
$$z = e^{xy} - \ln (1 + x^2 y^2)$$

The general form for a function of two variables is $z = f(x, y)$. Every function of two variables takes a Cartesian pair (x, y) as its input and in turn outputs a z value. Looking at the three previous examples, plugging in the input $(0, 1)$ gives you an output of 6 for the first function, 2 for the second function, and 1 for the third.

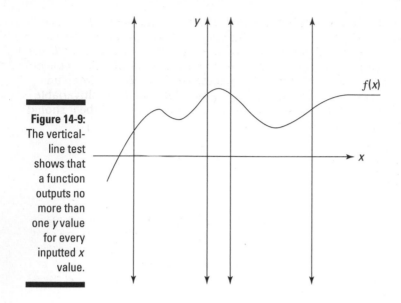

A good way to visualize a function of two variables is as a surface floating over the *xy*-plane in a 3-D Cartesian graph. (True, this surface can cross the plane and continue below it — just as a function can cross the *x*-axis — but for now, just picture it floating.) Every point on this surface looms directly above exactly one point on the plane. That is, if you pass a vertical line through any point on the plane, it crosses the function at no more than one point. Figure 14-10 illustrates this concept.

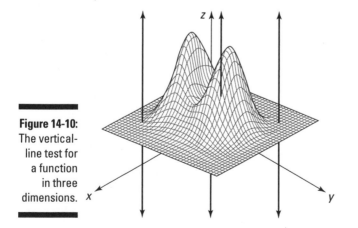

The concept of a function can be extended to higher dimensions. For example, $w = f(x, y, z)$ is the basic form for a function of three variables. Because this function (as well as other functions of more than two variables) exists in more than three dimensions, it's a lot harder to picture. For now, just concentrate on making the leap from two to three dimensions — that is, from functions of one variable to functions of two variables. Most of the multivariable calculus that you study in Calculus III is in three dimensions. (Maybe it should be called Calculus 3-D.)

Partial Derivatives

Partial derivatives are the higher-dimensional equivalent of the derivatives that you know from Calculus I. Just as a derivative represents the slope of a function on the Cartesian plane, a partial derivative represents a similar concept of slope in higher dimensions. In this section, I clarify this notion of slope in three dimensions. I also show you how to calculate the partial derivatives of functions of two variables.

Measuring slope in three dimensions

In the earlier section "Functions of Several Variables," I recommend that you visualize a function of two variables $z = f(x, y)$ as a surface floating over the xy-plane of a 3-D Cartesian graph. (See Figure 14-9 for a picture of a sample function.)

For example, take the function $z = y$, as shown in Figure 14-11. As you can see, this function looks a lot like the sloped roof of a house. Imagine yourself standing on this surface. When you walk parallel with the y-axis, your altitude either rises or falls. In other words, as the value of y changes, so does the value of z. But when you walk parallel with the x-axis, your altitude remains the same; changing the value of x has no effect on z.

So intuitively, you expect that the partial derivative $\frac{\partial z}{\partial y}$ — the slope in the direction of the y-axis — is 1. You also expect that the partial derivative $\frac{\partial z}{\partial x}$ — the slope in the direction of the x-axis — is 0. In the next section, I show you how to calculate partial derivatives to verify this result.

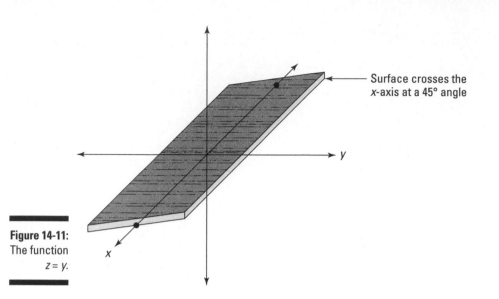

Surface crosses the
x-axis at a 45° angle

Figure 14-11:
The function
z = *y*.

Evaluating partial derivatives

Evaluating partial derivatives isn't much more difficult than evaluating regular derivatives. Given a function $z(x, y)$, the two partial derivatives are $\frac{\partial z}{\partial x}$ and $\frac{\partial z}{\partial y}$. Here's how you calculate them:

✔ To calculate $\frac{\partial z}{\partial x}$, treat y as a constant and use x as your differentiation variable.

✔ To calculate $\frac{\partial z}{\partial y}$, treat x as a constant and use y as your differentiation variable.

For example, suppose that you're given the equation $z = 5x^2y^3$. To find $\frac{\partial z}{\partial x}$, treat y as if it were a constant — that is, treat the entire factor $5y^3$ as if it's one big coefficient — and differentiate x^2:

$$\frac{\partial z}{\partial x} = 5y^3(2x) = 10xy^3$$

To find $\frac{\partial z}{\partial y}$, treat x as if it were a constant — that is, treat $5x^2$ as if it's the coefficient — and differentiate y^3:

$$\frac{\partial z}{\partial y} = 5x^2(3y^2) = 15x^2y^2$$

As another example, suppose that you're given the equation $z = 2e^x \sin y + \ln x$. To find $\frac{\partial z}{\partial x}$, treat y as if it were a constant and differentiate by the variable x:

$$\frac{\partial z}{\partial x} = 2 \sin y \, (e^x) + \frac{1}{x} = 2e^x \sin y + \frac{1}{x}$$

To find $\frac{\partial z}{\partial y}$, treat x as if it were a constant and differentiate by the variable y:

$$\frac{\partial z}{\partial y} = 2\,e^x \cos y$$

As you can see, when differentiating by y, the $\ln x$ term is treated as a constant and drops away completely.

Returning to the example from the previous section — the "sloped-roof" function $z = y$ — here are both partial derivatives of this function:

$$\frac{\partial z}{\partial x} = 0$$

$$\frac{\partial z}{\partial y} = 1$$

As you can see, this calculation produces the predicted results.

Multiple Integrals

You already know that an integral allows you to measure area in two dimensions (see Chapter 1 if this concept is unclear). And as you probably know from solid geometry, the analog of area in three dimensions is volume.

Multiple integrals are the higher-dimensional equivalent of the good old-fashioned integrals that you discover in Calculus II. They allow you to measure volume in three dimensions (or more).

Most of the multiple integrals that you'll ever have to solve come in two varieties: double integrals and triple integrals. In this section, I show you how to understand and calculate both of these types of multiple integrals.

Measuring volume under a surface

Definite integrals provide a reliable way to measure the signed area between a function and the x-axis as bounded by any two values of x. (I cover this in detail in Chapters 1 and 3.) Similarly, a *double integral* allows you to measure the signed volume between a function $z = f(x, y)$ and the xy-plane as bounded by any two values of x and any two values of y.

$$\int_0^2 \int_0^1 f(x,y)\,dx\,dy$$

To get a picture of this volume, look at Figure 14-12. The double integral measures the volume between $f(x, y)$ and the xy-plane as bounded by a rectangle. In this case, the rectangle is described by the four lines $x = 0$, $x = 1$, $y = 0$, and $y = 2$.

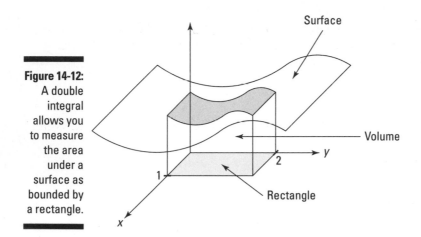

Figure 14-12:
A double integral allows you to measure the area under a surface as bounded by a rectangle.

A double integral is really an integral inside another integral. To help you see this, I bracket off the inner integral in the previous example:

$$\int_0^2 \left[\int_0^1 f(x,y)\, dx \right] dy$$

When you focus on the integral inside the brackets, you can see that the limits of integration 0 and 1 correspond with the dx — that is, $x = 0$ and $x = 1$. Similarly, the limits of integration 0 and 2 correspond with the dy — that is, $y = 0$ and $y = 2$.

Evaluating multiple integrals

Multiple integrals (double integrals, triple integrals, and so forth) are usually definite integrals, so evaluating them results in a real number. Evaluating multiple integrals is similar to evaluating nested functions: Work from the inside out.

Solving double integrals

Solve double integrals in two steps: First evaluate the inner integral, and then plug this solution into the outer integral and solve that. For example, suppose that you want to integrate the following double integral:

$$\int_1^3 \int_{4x}^{5x} \frac{y^2}{x}\, dy\ dx$$

To start out, place the inner integral in parentheses so that you can better see what you're working with:

$$\int_1^3 \left(\int_{4x}^{5x} \frac{y^2}{x} \, dy \right) dx$$

Now, focus on what's inside the parentheses. For the moment, you can ignore the rest. Your integration variable is y, so treat the variable x as if it were a constant, moving it outside the integral:

$$\frac{1}{x} \int_{4x}^{5x} y^2 \, dy$$

Notice that the limits of integration in this integral are functions of x. So the result of this definite integral will also be a function of x:

$$= \frac{y^3}{3x} \Big|_{y=4x}^{y=5x}$$

$$= \frac{5x^3}{3x} - \frac{4x^3}{3x} = \frac{125x^3}{3x} - \frac{64x^3}{3x} = \frac{61}{3} x^2$$

Now, plug this expression into the outer integral. In other words, substitute it for what's inside the parentheses:

$$\int_1^3 \frac{61}{3} x^2 \, dx$$

Evaluate this integral as usual:

$$= \frac{61}{9} x^3 \Big|_{x=1}^{x=3}$$

$$= \frac{61}{9} (3)^3 - \frac{61}{9} (1)^3$$

$$= 61 - \frac{61}{9} = \frac{488}{9}$$

Making sense of triple integrals

Triple integrals look scary, but if you take them step by step, they're no more difficult than regular integrals. As with double integrals, start in the center and work your way out. For example:

$$\int_1^2 \int_z^{4z} \int_{\frac{z}{y}}^{\frac{2z}{y}} x^3 y^5 z \, dx \, dy \, dz$$

Begin by separating the two inner integrals:

$$\int_1^2 \left[\int_z^{4z} \left(\int_{\frac{z}{y}}^{\frac{2z}{y}} x^3 y^5 z \, dx \right) dy \right] dz$$

Your plan of attack is to evaluate the integral in the brackets first, and then the integral in the braces, and finally the outer integral. First things first:

$$\int_{\frac{z}{y}}^{\frac{2z}{y}} x^3 y^5 z \, dx$$

$$= \frac{1}{4} x^4 y^5 z \Big|_{x=z/y}^{x=2z/y}$$

$$= \frac{1}{4} \left(\frac{2z}{y}\right)^4 y^5 z - \frac{1}{4} \left(\frac{z}{y}\right)^4 y^5 z$$

Notice that plugging in values for x results in an expression in terms of y and z. Now simplify this expression to make it easier to work with:

$$= 4yz^5 - \frac{1}{4} yz^5 = \frac{15}{4} yz^5$$

Plug this solution back in to replace the bracketed integral:

$$\int_{1}^{2} \left[\int_{z}^{4z} \frac{15}{4} yz^5 \, dy \right] dz$$

One integral down, two to go. This time, focus on the integral inside the braces. This time, the integration variable is y:

$$\int_{z}^{4z} \frac{15}{4} yz^5 \, dy$$

$$= \frac{15}{8} y^2 z^5 \Big|_{y=z}^{y=4z}$$

$$= \frac{15}{8} (4z)^2 z^5 - \frac{15}{8} z^2 z^5$$

Again, simplify before proceeding:

$$= 30z^7 - \frac{15}{8} z^7 = \frac{225}{8} z^7$$

Plug this result back into the integral as follows:

$$\int_{1}^{2} \frac{225}{8} z^7 \, dz$$

Now evaluate this integral:

$$= \frac{225}{64} z^8 \Big|_{z=1}^{z=2}$$

$$= \frac{225}{64} 2^8 - \frac{225}{64} 1^8$$

$$= \frac{57{,}600}{64} - \frac{225}{64} = \frac{57{,}375}{64}$$

Chapter 15

What's so Different about Differential Equations?

The very mention of differential equations (DEs for short) strikes a spicy combination of awe, horror, and utter confusion into nonmathematical minds. Even intrepid calculus students have been known to consider a career in art history when these untamed beasts come into focus on the radar screen. Just what are differential equations? Where do they come from? Why are they necessary? And how in the world do you solve them?

In this chapter, I answer these questions and give you some familiarity with DEs. I show you how to identify the basic types of DEs so that if you're ever at a math department cocktail party (lucky you!), you won't feel completely adrift. I relate DEs to the integrals that you now understand so well. I show you how to build your own DEs so that you'll always have a hobby to pass the time, and I also show you how to check DE solutions. In addition, you discover how DEs arise in physics. Finally, I show you a few simple methods for solving some basic differential equations.

Basics of Differential Equations

In a nutshell, a *differential equation* is any equation that includes at least one derivative. For example:

$$\frac{dy}{dx} = \sin x \, e^y$$

$$\frac{d^2 y}{dx^2} + 10\frac{dy}{dx} + 9y = 0$$

$$\frac{d^4 y}{dx^4} + \frac{d^3 y}{dx^3} + \frac{d^2 y}{dx^2} + \frac{dy}{dx} + y = \cos x$$

Solving a differential equation means finding the value of the dependent variable in terms of the independent variable. Throughout this chapter, I use y as the dependent variable, so the goal in each problem is to solve for y in terms of x.

In this section, I show you how to classify DEs. I also show you how to build DEs and check the solution to a DE.

Classifying DEs

As with other equations that you've encountered, differential equations come in many varieties. And different varieties of DEs can be solved by using different methods. In this section, I show you some important ways to classify DEs.

Ordinary and partial differential equations

An *ordinary differential equation* (ODE) has only derivatives of one variable — that is, it has no partial derivatives (flip to Chapter 14 for more on partial differentiation). Here are a few examples of ODEs:

$$\frac{dy}{dx} = x \sin (x^2) \cos y$$

$$\frac{dy}{dx} = y \csc x + e^x$$

$$\frac{d^2 y}{dx^2} + 4xy \frac{dy}{dx} + 5y = 0$$

In contrast, a *partial differential equation* (PDE) has at least one partial derivative. Here are a few examples of PDEs:

$$\frac{\partial u}{\partial x} + \frac{\partial u}{\partial y} + u = e^{x-y}$$

$$3\frac{\partial u}{\partial x^2} + 7\frac{\partial u}{\partial x \partial y} + 6\frac{\partial u}{\partial y^2} = 0$$

$$\frac{\partial v}{\partial t} - k\frac{\partial v}{\partial x^2}$$

Ordinary differential equations are usually the topic of a typical Differential Equations class in college. They're a step or two beyond what you're used to working with, but many students actually find Differential Equations an easier course than Calculus II (generally considered the most difficult class in the calculus series). However, ODEs are limited in how well they can actually express physical reality.

The real quarry is partial differential equations. A lot of physics gets done with these little gems. Unfortunately, solving PDEs is one giant leap forward in math from what the average calculus student is used to. Delving into the kind of math that makes PDEs come alive is typically reserved for graduate school.

Order of DEs

Differential equations are further classified according to their *order*. This classification is similar to the classification of polynomial equations by degree (see Chapter 2 for more on polynomials).

First-order ODEs contain only first derivatives. For example:

$$\frac{dy}{dx} = ye^x$$

$$3\frac{dy}{dx} = \sin y + 2e^{2x}$$

$$\ln xy\, \frac{dy}{dx} = 2x^2 + y - \tan x$$

Higher-order ODEs are classified, as polynomials are, by the *greatest* order of their derivatives. Here are examples of second-, third-, and fourth-order ODEs:

Second-order ODE: $\dfrac{d^2y}{dx^2} + 4\dfrac{dy}{dx} - 10y = e^x$

Third-order ODE: $\dfrac{d^3y}{dx^3} + \dfrac{d^2y}{dx^2} + y = 0$

Fourth-order ODE: $x^2\dfrac{d^4y}{dx^4} + \cos y = x$

As with polynomials, generally speaking, a higher-order DE is more difficult to solve than one of lower order.

Linear DEs

What constitutes a linear differential equation depends slightly on who you ask. For practical purposes, a linear first-order DE fits into the following form:

$$\frac{dy}{dx} + a(x)y = b(x)$$

where $a(x)$ and $b(x)$ are functions of x. Here are a few examples of linear first-order DEs:

$$\frac{dy}{dx} + y = x$$

$$\frac{dy}{dx} + 4xy = -\ln x$$

$$\frac{dy}{dx} - y \sin x = e^x$$

Linear DEs can often be solved, or at least simplified, using an *integrating factor*. I show you how to do this later in this chapter.

A linear second-degree DE fits into the following form:

$$a\frac{d^2 y}{dx^2} + b\frac{dy}{dx} + cy = 0$$

where a, b, and c are all constants. Here are some examples:

$$\frac{d^2 y}{dx^2} + 3\frac{dy}{dx} + 4y = 0$$

$$2\frac{d^2 y}{dx^2} + 5\frac{dy}{dx} + 6y = 0$$

$$\frac{d^2 y}{dx^2} = 0$$

Note that the constant a can always be reduced to 1, resulting in adjustments to the other two coefficients. Linear second-degree DEs are usually an important topic in a college-level course in differential equations. Solving them requires knowledge of matrices and complex numbers that is beyond the scope of this book.

Looking more closely at DEs

You don't have to play professional baseball to enjoy baseball. Instead, you can enjoy the game from the bleachers or, if you prefer, from a nice cushy chair in front of the TV. Similarly, you don't have to get too deep into differential equations to gain a general understanding of how they work. In this section, I give you front-row seats to the game of differential equations.

How every integral is a DE

The integral is a particular example of a more general type of equation — the differential equation. To see how this is so, suppose that you're working with this nice little integral:

$$y = \int \cos x \ dx$$

Differentiating both sides turns it into a DE:

$$\frac{dy}{dx} = \cos x$$

Of course, you know how to solve this DE by thinking of it as an integral:

$$y = \sin x + C$$

So in general, when a DE is of the form

$$\frac{dy}{dx} = f(x)$$

with $f(x)$ an arbitrary function of x, you can express that DE as an integral and solve it by integrating.

Why building DEs is easier than solving them

The reason that the DE in the last section is so simple to solve is that the derivative is isolated on one side of the equation. DEs attain a new level of difficulty when the derivative isn't isolated.

A good analogy can be made in lower math, when you make the jump from arithmetic to algebra. For example, here's an arithmetic problem:

$$x = 20 - (4^2 + 3)$$

Even though this is technically an algebra problem, you can solve it without algebra because x is isolated at the start of the problem. However, the ball-game changes when x becomes more enmeshed in the equation. For example:

$$2x^3 - x^2 + 5x - 17 = 0$$

Arithmetic isn't strong enough for this problem, so algebra takes over. Similarly, when derivatives get entangled into the fabric of an equation — as in most of the DEs I show you earlier in "Classifying DEs" — integrating is no longer effective and the search for new methods begins.

Although solving DEs is often tricky, building them is easy. For example, suppose that you start with this simple quadratic equation:

$$y = 3x^2 + 4x - 5$$

Now, find the first and second derivatives:

$$\frac{dy}{dx} = 6x + 4$$

$$\frac{d^2 y}{dx^2} = 6$$

Adding up the left and right sides of all three equations gives you the following differential equation:

$$y + \frac{dy}{dx} + \frac{d^2y}{dx^2} = 3x^2 + 10x + 5$$

Because you built the equation yourself, you know what y equals. But if you hand this equation off to some other students, they probably wouldn't be able to guess how you built it, so they would have to do some work to solve it for y. For example, here's another DE:

$$y + \frac{dy}{dx} + \frac{d^2y}{dx^2} + \frac{d^2y}{dx^3} = 0$$

This equation probably looks difficult because you don't have much information. And yet, after I tell you the solution, it appears simple:

$$y = \sin x \qquad \frac{dy}{dx} = \cos x$$

$$\frac{d^2y}{dx^2} = -\sin x \qquad \frac{d^3y}{dx^4} = -\cos x$$

But even after you have the solution, how do you know whether it's the *only* solution? For starters, $y = -\sin x$, $y = \cos x$, and $y = -\cos x$ are all solutions. Do other solutions exist? How do you find them? And how do you know when you have them all?

Another difficulty arises when y itself becomes tangled up in the equation. For example, how do you solve this equation for y?

$$\frac{dy}{dx} = \frac{\sin x}{e^y}$$

As you can see, differential equations contain treacherous subtleties that you don't find in basic calculus.

Checking DE solutions

Even if you don't know how to find a solution to a differential equation, you can always check whether a proposed solution works. This is simply a matter of plugging the proposed value of the dependent variable — I use y throughout this chapter — into both sides of the equation to see whether equality is maintained.

For example, here's a DE:

$$\frac{dy}{dx} = 3y + 4e^{3x} \cos x$$

You may not have a clue how to begin solving this DE, but imagine that an angel lands on your pen and offers you this solution:

$$y = 4e^{3x} \sin x$$

You can check to see whether this angel really knows math by plugging in this value of y as follows:

$$\frac{dy}{dx} = 3y + 4e^{3x} \cos x$$

$$\frac{d}{dx} \, 4e^{3x} \sin x = 3(4e^{3x} \sin x) + 4e^{3x} \cos x$$

$$4(3e^{3x} \sin x + e^{3x} \cos x) = 12 \, e^{3x} \sin x + 4e^{3x} \cos x$$

$$= 12 \, e^{3x} \sin x + 4e^{3x} \cos x = 12 \, e^{3x} \sin x + 4e^{3x} \cos x$$

Because the left and right sides of the equation are equal, the angel's solution checks out.

Solving Differential Equations

In this section, I show you how to solve a few types of DEs. First, you solve everybody's favorite DE, the *separable equation*. Next, you put this understanding to work to solve an *initial-value problem* (IVP). Finally, I show you how to solve a linear first-order DE by using an integrating factor.

Solving separable equations

Differential equations become harder to solve the more entangled they become. In certain cases, however, an equation that looks all tangled up is actually easy to tease apart. Equations of this kind are called *separable equations* (or *autonomous equations*), and they fit into the following form:

$$\frac{dy}{dx} = f(x) \cdot g(y)$$

Separable equations are relatively easy to solve. For example, suppose that you want to solve the following problem:

$$\frac{dy}{dx} = \frac{\sin x}{e^y}$$

You can think of the symbol $\frac{dy}{dx}$ as a fraction and isolate the x and y terms of this equation on opposite sides of the equal sign:

$$e^y \, dy = \sin x \, dx$$

Now, integrate both sides:

$$\int e^y \, dy = \int \sin x \, dx$$

$$e^y + C_1 = -\cos x + C_2$$

In an important sense, the previous step is questionable because the variable of integration is different on each side of the equal sign. You may think "No problem, it's all integration!" But imagine if you tried to divide one side of an equation by 2 and the other by 3, and then laughed it off with "It's all division!" Clearly, you'd have a problem. The good news, however, is that for technical reasons beyond the scope of this book, integrating both sides by different variables actually produces the correct answer.

C_1 and C_2 are both constants, so you can use the equation $C = C_2 - C_1$ to simplify the equation:

$$e^y = -\cos x + C$$

Next, use a natural log to undo the exponent, and then simplify:

$$\ln e^y = \ln (-\cos x + C)$$
$$y = \ln (-\cos x + C)$$

To check this solution, substitute this value for y into both sides of the original equation:

$$\frac{dy}{dx} = \frac{\sin x}{e^y}$$

$$\frac{d}{dx} \ln(-\cos x + C) = \frac{\sin x}{e^{\ln(-\cos x = C)}}$$

$$\frac{d}{dx} \ln(-\cos x + C) = \frac{\sin x}{-\cos x + C}$$

$$\frac{1}{-\cos x + C} \cdot \sin x = \frac{\sin x}{-\cos x + C}$$

$$\frac{\sin x}{-\cos x + C} = \frac{\sin x}{-\cos x + C}$$

Solving initial-value problems (IVPs)

In Chapter 3, I show you that the definite integral is a particular example of a whole family of indefinite integrals. In a similar way, an *initial-value problem* (IVP) is a particular example of a solution to a differential equation. Every IVP gives you extra information — called an *initial value* — that allows you to use the *general solution* to a DE to obtain a *particular solution*.

For example, here's an initial-value problem:

$$\frac{dy}{dx} = y \sec^2 x \quad y(0) = 5$$

This problem includes not only a DE, but also an additional equation. To understand what this equation tells you, remember that y is a dependent variable, a function of x. So, the notation $y(0) = 5$ means "when $x = 0$, $y = 5$." You see how this information comes into play as I continue with this example.

To solve an IVP, you first have to solve the DE. Do this by finding its general solution without worrying about the initial value. Fortunately, this DE is a separable equation, which you know how to solve from the last section:

$$\frac{1}{y} \, dy = \sec^2 x \, dx$$

Integrate both sides:

$$\int \frac{1}{y} \, dy = \int \sec^2 x \, dx$$

$$\ln y = \tan x + C$$

In this last step, I use C to consolidate the constants of integration from both sides of the equation into a single constant C. (If this doesn't make sense, I explain why in "Solving separable equations" earlier in this chapter.) Next, I undo the natural log by using e:

$$e^{\ln y} = e^{\tan x + C}$$

$$y = e^{\tan x} \cdot e^{C}$$

Because e^{C} is a constant, this equation can be further simplified by using the substitution $D = e^{C}$:

$$y = De^{\tan x}$$

Before moving on, check to make sure that this solution is correct by substituting this value of y into both sides of the original equation:

$$\frac{dy}{dx} = y \sec^2 x$$

$$\frac{d}{dx} De^{\tan x} = De^{\tan x} \sec^2 x$$

$$De^{\tan x} \sec^2 x = De^{\tan x} \sec^2 x$$

This checks out, so $y = De^{\tan x}$ is, indeed, the *general solution* to the DE. To solve the initial-value problem, however, I need to find the specific value of the variable D by using the additional information I have: When $x = 0$, $y = 5$. Plugging both of these values into the equation makes it possible to solve for D:

$$5 = De^{\tan 0}$$

$$5 = De^0$$

$$5 = D$$

Now, plug this value of D back into the general solution of the problem to get the IVP solution:

$$y = 5e^{\tan x}$$

This solution satisfies not only the differential equation $\dfrac{dy}{dx} = y \sec^2 x$ but also the initial value $y(0) = 5$.

Using an integrating factor

As I mention earlier in this chapter, in "Classifying DEs," a linear first-order equation takes the following form:

$$\frac{dy}{dx} + a(x)y = b(x)$$

A clever method for solving DEs in this form involves multiplying the entire equation by an *integrating factor*. Follow these steps:

1. **Calculate the integrating factor.**

2. **Multiply the DE by this integrating factor.**

3. **Restate the left side of the equation as a single derivative.**

4. **Integrate both sides of the equation and solve for *y*.**

Don't worry if these steps don't mean much to you. In the upcoming sections, I show you what an integrating factor is and how to use it to solve linear first-order DEs.

Getting very lucky

To help you understand how multiplying by an integrating factor works, I set up an equation that practically solves itself — that is, if you know what to do:

$$\frac{dy}{dx} + \frac{2y}{x} = 0$$

Notice that this is a linear first-degree DE, with $a(x) = \frac{2}{x}$ and $b(x) = 0$. I now tweak this equation by multiplying every term by x^2 (you see why shortly):

$$\frac{dy}{dx} \cdot x^2 + \frac{2y}{x} \cdot x^2 = 0 \cdot x^2$$

Next, I use algebra to do a little simplifying and rearranging:

$$\frac{dy}{dx} \cdot x^2 + 2x \cdot y = 0$$

Now, here's where I appear to get extremely lucky: The two terms on the left side of the equation just happen to be the *result* of the application of the Product Rule to the expression $y \cdot x^2$ (for more on the Product Rule, see Chapter 2):

$$\frac{d}{dx}(y \cdot x^2) = \frac{dy}{dx} \cdot x^2 + 2x \cdot y$$

Notice that the right side of this equation is exactly the same as the left side of the previous equation. So I can make the following substitution:

$$\frac{d}{dx}(y \cdot x^2) = 0$$

Now, to undo the derivative on the left side, I integrate both sides, and then I solve for y:

$$\int \left[\frac{d}{dx}(y \cdot x^2) \right] dx = \int 0 \; dx$$

$$y \cdot x^2 = C$$

$$y = \frac{C}{x^2}$$

To check this solution, I plug this value of y back into the original equation:

$$\frac{dy}{dx} + \frac{2y}{x} = 0$$

$$\frac{d}{dx} \frac{C}{x^2} + \frac{2}{x} \cdot \frac{C}{x^2} = 0$$

$$\frac{-2C}{x^3} + \frac{2C}{x^3} = 0$$

$$0 = 0$$

Making your luck

The previous example works because I found a way to multiply the entire equation by a factor that made the left side of the equation look like a derivative resulting from the Product Rule. Although this looked lucky, if you know what to multiply by, *every* linear first-order DE can be transformed in this way. Recall that the form of a linear first-order DE is as follows:

$$\frac{dy}{dx} + a(x)y = b(x)$$

The trick is to multiply the DE by an *integrating factor* based on $a(x)$. Here's the integrating factor:

$$e^{\int a\,(x)\,dx}$$

How the integrating factor is originally derived is beyond the scope of this book. All you need to know here is that it works.

For example, in the previous problem, you know that $a(x) = \frac{2}{x}$. So here's how to find the integrating factor:

$$e^{\int a\,(x)\,dx} = e^{\int \frac{2}{x}\,dx} = e^{2\ln x}$$

Remember that $2 \ln x = \ln x^2$, so:

$$e^{\ln x^2}$$

$$= x^2$$

As you can see, the integrating factor x^2 is the exact value that I multiplied by to solve the problem. To see how this process works now that you know the trick, here's another DE to solve:

$$\frac{dy}{dx} + 3y = e^x$$

In this case, $a(x) = 3$, so compute the integrating factor as follows:

$$e^{\int a\,(x)\,dx} = e^{\int 3\,dx} = e^{3x}$$

Now, multiply every term in the equation by this factor:

$$\frac{dy}{dx} \cdot e^{3x} + 3y \cdot e^{3x} = e^x \cdot e^{3x}$$

If you like, use algebra to simplify the right side and rearrange the left side:

$$\frac{dy}{dx} \cdot e^{3x} + 3e^{3x} \cdot y = e^{4x}$$

Now, you can see how the left side of this equation looks like the *result* of the Product Rule applied to evaluate the following derivative:

$$\frac{d}{dx}(y \cdot e^{3x}) = \frac{dy}{dx} \cdot e^{3x} + 3e^{3x} \cdot y$$

Because the right side of this equation is the same as the left side of the previous equation, I can make the following substitution:

$$\frac{d}{dx}(y \cdot e^{3x}) = e^{4x}$$

Notice that I change the left side of the equation by using the Product Rule *in reverse*. That is, I'm expressing the whole left side as a single derivative. Now, I can integrate both sides to undo this derivative:

$$\int \left[\frac{d}{dx}(y \cdot e^{3x})\right] dx = \int e^{4x} \, dx$$

$$y \cdot e^{3x} = \frac{e^{4x}}{4} + C$$

Now, solve for *y* and simplify:

$$y = \frac{e^{4x}}{4e^{3x}} + \frac{C}{e^{3x}} = \frac{e^{x}}{4} + Ce^{-3x}$$

To check this answer, substitute this value of *y* back into the original DE:

$$\frac{dy}{dx} + 3y = e^{x}$$

$$\frac{d}{dx}\left(\frac{e^{x}}{4} + Ce^{-3x}\right) + 3\left(\frac{e^{x}}{4} + Ce^{-3x}\right) = e^{x}$$

$$\frac{e^{x}}{4} - 3Ce^{-3x} + \frac{3e^{x}}{4} + 3Ce^{-3x} = e^{x}$$

$$\frac{e^{x}}{4} + \frac{3e^{x}}{4} = e^{x}$$

$$e^{x} = e^{x}$$

As if by magic, this answer checks out, so the solution is valid.

Part VI
The Part of Tens

The 5th Wave   By Rich Tennant

CALCULUS TEST TODAY!!

"Watch this, Ruth. Steady ... steady ... calculate the volume of the stein first."

In this part . . .

As a special bonus, here are two top-ten lists on calculus-related topics: ten important Calculus II "aha" moments and ten useful test-taking tips.

Chapter 16

Ten "Aha!" Insights in Calculus II

In This Chapter

▶ Understanding the key concepts of integration

▶ Distinguishing the definite integral from the indefinite integral

▶ Knowing the basics of infinite series

*O*kay, here you are near the end of the book. You read every single word that I wrote, memorized the key formulas, and worked through all the problems. You're all set to ace your final exam, and you've earned it. Good for you! (Or maybe you just picked up the book and skipped to the end. That's fine, too! This is a great place to get an overview of what this Calculus II stuff is all about.)

But still, you have this sneaking suspicion that you're stuck in the middle of the forest and can't see it because of all those darn trees. Forget the equations for a moment and spend five minutes looking over these top ten "Aha!" insights. When you understand them, you have a solid conceptual framework for Calculus II.

Integrating Means Finding the Area

Finding the area of a polygon or circle is easy. Integration is all about finding the area of shapes with weird edges that are hard to work with. These edges may be the curves that result from polynomials, exponents, logarithms, trig functions, or inverse trig functions, or the products and compositions of these functions.

Integration gives you a concrete way to look at this question, known as the *area problem*. No matter how complicated integration gets, you can always understand what you're working on in terms of this simple question: "How does what I'm doing help me find an area?"

See Chapter 1 for more about the relationship between integration and area.

When You Integrate, Area Means Signed Area

In the real world, area is always positive. For example, there's no such thing as a piece of land that's –4 square miles in area. This concept of area is called *unsigned area*.

But on the Cartesian graph in the context of integration, area is measured as *signed area*, with area below the *x*-axis considered to be *negative area*.

In this context, a 2-x-2-unit square below the *x*-axis is considered to be –4 square units in signed area. Similarly, a 2-x-2-unit square that's divided in half by the *x*-axis is considered to have an area of 0.

The definite integral always produces the signed area between a curve and the *x*-axis, within the limits of integration. So if an application calls for the unsigned area, you need to measure the positive area and negative area separately, change the sign of the negative area, and add these two results together.

See Chapter 3 for more about signed area.

Integrating Is Just Fancy Addition

To measure the area of an irregularly shaped polygon, a good first step is to cut it into smaller shapes that you know how to measure — for example, triangles and rectangles — and then add up the areas of these shapes.

Integration works on the same principle. It allows you to slice a shape into smaller shapes that approximate the area that you're trying to measure, and then add up the pieces. In fact, the integral sign $\int$ itself is simply an elongated *S*, which stands for *sum*.

See Chapter 1 for more about how integration relates to addition.

Integration Uses Infinitely Many Infinitely Thin Slices

Here's where integration differs from other methods of measuring area: Integration allows you to slice an area into infinitely many pieces, all of which are infinitely thin, and then add up these pieces to find the total area.

Or, to put a slightly more mathematical spin on it: The definite integral is the limit of the total area of all these slices as the number of slices approaches infinity and the thickness of each slice approaches 0.

This concept is also useful when you're trying to find volume, as I show you in Chapter 10.

See Chapter 1 for more about how this concept of infinite slicing relates to integration.

Integration Contains a Slack Factor

Math is a harsh mistress. A small error at the beginning of a problem often leads to a big mistake by the end.

So finding out that you can thin-slice an area in a bunch of different ways and still get the correct answer is refreshing. Some of these methods for thin-slicing include left rectangles, right rectangles, and the midpoint rectangles. I cover them all in Chapter 3.

This *slack factor,* as I call it, comes about because integration exploits an infinite sequence of successive approximations. Each approximation brings you closer to the answer that you're seeking. So, no matter what route you take to get there, an infinite number of such approximations brings you to the answer.

See Chapter 3 for more about the distinction between approximating and evaluating integrals.

A Definite Integral Evaluates to a Number

A definite integral represents the well-defined area of a shape on a graph. You can represent any such area as a number of square units, so the definite integral is a number.

See Chapter 3 for more about the definite integral.

An Indefinite Integral Evaluates to a Function

An indefinite integral is a template that allows you to calculate an infinite number of related definite integrals by plugging in some parameters. In math, such a template is called a *function*.

The input values to an indefinite integral are the two limits of integration. Specifying these two values turns the indefinite integral into a definite integral, which then outputs a number representing an area.

But if you don't specify the limits of integration, you can still evaluate an indefinite integral as a function. The process of finding an indefinite integral turns an input function (for example, cos x) into an output function (sin $x + C$).

See Chapter 3 for more about the indefinite integral and Part II for a variety of techniques for evaluating indefinite integrals.

Integration Is Inverse Differentiation

Integration and differentiation are inverse operations: Either of these operations undoes the other (up to a constant C). Another way to say this is that integration is *anti-differentiation*.

Here's an example of how differentiation undoes integration:

$$\int 5x^3\,dx = \frac{5}{4}x^4 + C$$

$$\frac{d}{dx}\,\frac{5}{4}\,x^4 + C = 5x^3$$

As you can see, integrating a function and then differentiating the result produces the function that you started with.

Now, here's an example of how integration undoes differentiation:

$$\frac{d}{dx}\,\sin x = \cos x$$

$$\int \cos x\,dx = \sin x + C$$

As you can see, differentiating a function and then integrating the result produces the function that you started with, plus a constant C.

See Part II for more on how this inverse relationship between integration and differentiation provides a variety of clever methods for integrating complicated functions.

Every Infinite Series Has Two Related Sequences

Every infinite series has two related sequences that are important for understanding how that series works: its defining sequence and its sequence of partial sums.

The *defining sequence* of a series is simply the sequence that defines the series in the first place. For example, the series

$$\sum_{n=1}^{\infty} \frac{1}{n} = 1 + \frac{1}{2} + \frac{1}{3} + \frac{1}{4} + \frac{1}{5} + \dots$$

has the defining sequence

$$\left\{ \frac{1}{n} \right\} = \left\{ 1, \frac{1}{2}, \frac{1}{3}, \frac{1}{4}, \frac{1}{5} \dots \right\}$$

Notice that the same function — in this case, $\frac{1}{n}$ — appears in the shorter notation for both the series and its defining sequence.

The *sequence of partial sums* of a series is the sequence that results when you successively add a finite number of terms. For example, the previous series has the following sequence of partial sums:

$$1, \frac{3}{2}, \frac{11}{6}, \frac{25}{12}, \frac{137}{60}, \dots$$

Notice that a series may diverge while its defining sequence converges, as in this example. However, a series and its sequence of partial sums always converge or diverge together. In fact, the definition of convergence for a series is based upon the behavior of its sequence of partial sums (see the next section for more on convergence and divergence).

See Part IV for more about infinite series.

Every Infinite Series Either Converges or Diverges

Every infinite series either converges or diverges, with no exceptions.

A series *converges* when it evaluates to (equals) a real number. For example:

$$\sum_{n=1}^{\infty} \left(\frac{1}{2}\right)^{n} = \frac{1}{2} + \frac{1}{4} + \frac{1}{8} + \frac{1}{16} + \ldots = 1$$

On the other hand, a series *diverges* when it doesn't evaluate to a real number. Divergence can happen in two different ways. The more common type of divergence is when the series explodes to ∞ or $-\infty$. For example:

$$\sum_{n=1}^{\infty} n = 1 + 2 + 3 + 4 + \ldots$$

Clearly, this series doesn't add up to a real number — it just keeps getting bigger and bigger forever.

Another type of divergence occurs when a series bounces forever among two or more values. This happens only when a series is *alternating* (see Chapter 12 for more on alternating series). For example:

$$\sum_{n=1}^{\infty} (-1)^{n} = -1 + 1 - 1 + 1 - \ldots$$

The sequence of partial sums (see the previous section) for this series alternates forever between -1 and 0, never settling in to a single value, so the series diverges.

See Part IV for more about infinite series.

Chapter 17

Ten Tips to Take to the Test

. . . .

In This Chapter

▶ Staying calm when the test is passed out

▶ Remembering those dxs and $+ C$s

▶ Getting unstuck

▶ Checking for mistakes

. . . .

I've never met anyone who loved taking a math test. The pressure is on, the time is short, and that formula that you can't quite remember is out of reach. Unfortunately, exams are a part of every student's life. Here are my top ten suggestions to make test-taking just a little bit easier.

Breathe

This is always good advice — after all, where would you be if you weren't breathing? Well, not a very nice place at all.

A lot of what you may feel when facing a test — for example, butterflies in your stomach, sweaty palms, or trembling — is simply a physical reaction to stress that's caused by adrenalin. Your body is preparing you for a fight-or-flight response, but with a test, you have nothing to fight and nowhere to fly.

A little deep breathing is a simple physical exertion that can help dissipate the adrenalin and calm you down. So, while you're waiting for the professor to arrive and hand out the exams, take a few deep breaths in and out. If you like, picture serenity and deep knowledge of all things mathematical entering your body on the in-breath, and all the bad stuff exiting on the out-breath.

Start by Reading through the Exam

When you receive your exam, take a minute to read through it so that you know what you're up against. This practice starts your brain working (consciously or not) on the problems.

While you're reading, see whether you can find a problem that looks easier to you than the others (see the next section).

Solve the Easiest Problem First

After the initial read-through, turn to the page with the easiest problem and solve it. This warm-up gets your brain working and usually reduces your anxiety.

Don't Forget to Write dx and + C

Remember to include those pesky little *dx*s in every integration statement. They need to be there, and some professors take it very personally when you don't include them. You have absolutely no reason to lose points over something so trivial.

And don't forget that the solution to every *indefinite* integral ends with + C (or whatever constant you choose). No exceptions! As with the *dx*s, omitting this constant can cost you points on an exam, so get in the habit of including it.

Take the Easy Way Out Whenever Possible

In Chapters 4 through 8, I introduce the integration techniques in the order of difficulty. Before you jump in to your calculation, take a moment to walk through all the methods you know, from easiest to hardest.

Always check first to see whether you know a simple formula: For example, $\int \frac{1}{x\sqrt{x^2-1}}\,dx$ may cause you to panic until you remember that the answer is simply arcsec x + C. If no formula exists, think through whether a simple variable substitution is possible. What about integration by parts? Your last resorts are always trig substitution and integration with partial fractions.

When you're working on solving area problems, stay open to the possibility that calculus may not be necessary. For example, you don't need calculus to find the area under a straight line or semicircle. So, before you start integrating, step back for a moment to see whether you can spot an easier way.

If You Get Stuck, Scribble

When you look at a problem and you just don't know which way to go, grab a piece of scratch paper and scribble everything you can think of, without trying to make sense of it.

Use algebra, trig identities, and variable substitutions of all kinds. Write series in both sigma notation and expanded notation. Draw pictures and graphs. Write it all down, even the ideas that seem worthless.

You may find that this process jogs your brain. Even copying the problem — equations, graphs, and all — can sometimes help you to notice something important that you missed in your first reading of the question.

If You Really Get Stuck, Move On

I see no sense in beating your head against a brick wall, unless you like getting brick dust in your hair. Likewise, I see no sense in spending the whole exam frozen in front of one problem.

So, after you scribble and scribble some more (see the previous section) and you're still getting nowhere with a problem, move on. You may as well make the most of the time you're given by solving the problems that you can solve. What's more, many problems seem easier on the second try. And working on other areas of the test may remind you of some important information that you'd forgotten.

Check Your Answers

Toward the end of the test, especially if you're stuck, take a moment to check over some of the problems that you already completed. Does what you've written still make sense? If you see any missing dxs or + Cs, fill them in. Make sure you didn't drop any minus signs. Most important, do a reality check of your answer compared with the original problem to see whether it makes sense.

For example, suppose that you're integrating to find an area someplace inside a 2 x 2 region on a graph, and your answer is 7 trillion. Obviously, something went wrong. If you have time to find out what happened, trace back over your steps.

Although fixing a problem on an exam can be tedious, it usually takes less time than starting (and maybe not finishing) a brand-new problem from scratch.

If an Answer Doesn't Make Sense, Acknowledge It

Suppose that you're integrating to find an area someplace inside a 2 x 2 region on a graph, and your answer is 7 trillion. Obviously, something went wrong. If you don't have time to find out what happened, write a note to the professor acknowledging the problem.

Writing such a note lets your professor know that your conceptual understanding of the problem is okay — that is, you get the idea that integration means area. So, if it turns out that your calculation got messed up because of a minor mistake like a lost decimal point, you'll probably lose only a couple of points.

Repeat the Mantra "I'm Doing My Best," and Then Do Your Best

All you can do is your best, and even the best math student occasionally forgets a formula or stares at an exam question and goes "Huh?"

When these moments arrive, and they will, you can do a shame spiral about all the studying you shoulda, coulda, woulda done. But there's no cheese down that tunnel. You can also drop your pencil, leave the room, quit school, fly to Tibet, and join a monastery. This plan of action is also not recommended unless you're fluent in Tibetan (which is way harder than calculus!).

Instead, breathe (see the section on breathing earlier in this chapter) and gently remind yourself "I'm doing my best." And then do your best with what you have. Perfection is not of this world, but if you can cut yourself a bit of slack when you're under pressure, you'll probably end up doing better than you would've otherwise.

Index

• S •

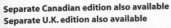

SPORTS, FITNESS, PARENTING, RELIGION & SPIRITUALITY

0-471-76871-5

0-7645-7841-3

Also available:

- Catholicism For Dummies
  0-7645-5391-7
- Exercise Balls For Dummies
  0-7645-5623-1
- Fitness For Dummies
  0-7645-7851-0
- Football For Dummies
  0-7645-3936-1
- Judaism For Dummies
  0-7645-5299-6
- Potty Training For Dummies
  0-7645-5417-4
- Buddhism For Dummies
  0-7645-5359-3

- Pregnancy For Dummies
  0-7645-4483-7 †
- Ten Minute Tone-Ups For Dummies
  0-7645-7207-5
- NASCAR For Dummies
  0-7645-7681-X
- Religion For Dummies
  0-7645-5264-3
- Soccer For Dummies
  0-7645-5229-5
- Women in the Bible For Dummies
  0-7645-8475-8

TRAVEL

0-7645-7749-2

0-7645-6945-7

Also available:

- Alaska For Dummies
  0-7645-7746-8
- Cruise Vacations For Dummies
  0-7645-6941-4
- England For Dummies
  0-7645-4276-1
- Europe For Dummies
  0-7645-7529-5
- Germany For Dummies
  0-7645-7823-5
- Hawaii For Dummies
  0-7645-7402-7

- Italy For Dummies
  0-7645-7386-1
- Las Vegas For Dummies
  0-7645-7382-9
- London For Dummies
  0-7645-4277-X
- Paris For Dummies
  0-7645-7630-5
- RV Vacations For Dummies
  0-7645-4442-X
- Walt Disney World & Orlando
  For Dummies
  0-7645-9660-8

GRAPHICS, DESIGN & WEB DEVELOPMENT

0-7645-8815-X

0-7645-9571-7

Also available:

- 3D Game Animation For Dummies
  0-7645-8789-7
- AutoCAD 2006 For Dummies
  0-7645-8925-3
- Building a Web Site For Dummies
  0-7645-7144-3
- Creating Web Pages For Dummies
  0-470-08030-2
- Creating Web Pages All-in-One Desk
  Reference For Dummies
  0-7645-4345-8
- Dreamweaver 8 For Dummies
  0-7645-9649-7

- InDesign CS2 For Dummies
  0-7645-9572-5
- Macromedia Flash 8 For Dummies
  0-7645-9691-8
- Photoshop CS2 and Digital
  Photography For Dummies
  0-7645-9580-6
- Photoshop Elements 4 For Dummies
  0-471-77483-9
- Syndicating Web Sites with RSS Feeds
  For Dummies
  0-7645-8848-6
- Yahoo! SiteBuilder For Dummies
  0-7645-9800-7

NETWORKING, SECURITY, PROGRAMMING & DATABASES

0-7645-7728-X

0-471-74940-0

Also available:

- Access 2007 For Dummies
  0-470-04612-0
- ASP.NET 2 For Dummies
  0-7645-7907-X
- C# 2005 For Dummies
  0-7645-9704-3
- Hacking For Dummies
  0-470-05235-X
- Hacking Wireless Networks
  For Dummies
  0-7645-9730-2
- Java For Dummies
  0-470-08716-1

- Microsoft SQL Server 2005 For Dummies
  0-7645-7755-7
- Networking All-in-One Desk Reference
  For Dummies
  0-7645-9939-9
- Preventing Identity Theft For Dummies
  0-7645-7336-5
- Telecom For Dummies
  0-471-77085-X
- Visual Studio 2005 All-in-One Desk
  Reference For Dummies
  0-7645-9775-2
- XML For Dummies
  0-7645-8845-1